Introduction to crystal chemistry

Student edition

Introduction to crystal chemistry

Student edition

HOWARD W. JAFFE

University of Massachusetts at Amherst

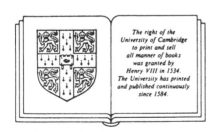

The right of the
University of Cambridge
to print and sell
all manner of books
was granted by
Henry VIII in 1534.
The University has printed
and published continuously
since 1584.

Cambridge University Press

Cambridge

New York   Port Chester   Melbourne   Sydney

CAMBRIDGE UNIVERSITY PRESS
Cambridge, New York, Melbourne, Madrid, Cape Town, Singapore, São Paulo, Delhi

Cambridge University Press
The Edinburgh Building, Cambridge CB2 8RU, UK

Published in the United States of America by Cambridge University Press, New York

www.cambridge.org
Information on this title: www.cambridge.org/9780521369855

First published 1988
Reprinted 1990
Re-issued in this digitally printed version 2009

A catalogue record for this publication is available from the British Library

Library of Congress Cataloguing in Publication data
Jaffe, Howard W.
[Crystal chemistry and refractivity, Part 1]
Introduction to crystal chemistry / Howard W. Jaffe, – Student ed.
    p.   cm.
"This student edition . . . constitutes Part 1 (Chapters 1–11) of a
larger two-part monograph volume, *Crystal chemistry and
refractivity*" – Pref.
Includes index.
1. Crystallography.   I. Title.
QD921.J283   1988
548 – dc19   88 – 6517

ISBN 978-0-521-36985-5 paperback

To
Elizabeth Boudreau Jaffe

Contents

Preface to the student edition

This book covers material taught in the Department of Geology and Geography, at the University of Massachusetts, Amherst, in courses offered to advanced undergraduate and beginning graduate students who have completed courses in introductory mineralogy and chemistry. It is intended for use as a textbook for any second course in mineralogy, or for any introductory course in crystal chemistry or the materials sciences.

Because crystal chemistry embraces the constitution of all forms of matter, it must draw upon and relate data obtained from a variety of scientific disciplines. This book *integrates* essential data on atomic and electronic structure, orbitals and chemical bonding, packing and coordination of atoms, polymerization and distortion of polyhedra, crystal field theory, cation ordering, and the relation of optical properties to density and chemical composition imposed by crystal structure. The current search for new crystalline materials for use in lasers, superconductors, and other applications has only begun, and offers many challenges and potential rewards to the astute student of crystal chemistry.

This student edition, *Introduction to Crystal Chemistry,* constitutes Part I (Chapters 1–11: "Principles of Crystal Chemistry and Refractivity") of a larger two-part monograph volume, *Crystal Chemistry and Refractivity.* The several references to Chapters 12–20 contained in the student edition (or to pp. 147–325 in the indexes) refer to those chapters comprising Part II ("Descriptive Crystal Chemistry") of the larger work.

This introductory edition is intended to give the student the background necessary to understand the organization of atomic arrays that form the complex silicate, oxide, borate, carbonate, sulfate, sulfide, and other crystal structures treated in Part II.

Amherst, Mass.
January 1988

H. W. J.

Acknowledgments

The author wishes to thank The Institute of Mineralogy, Section of Earth Sciences, University of Geneva, Switzerland, and its former head, Professor Marc B. Vuagnat, for the use of its superb international mineralogical library and other facilities.

Margarita Shannon and David Elbert worked out the mathematical formula for the direct calculation of the unit cell volume of a trigonal–rhombohedral mineral, using the cell edge, a_{rh}, and the interaxial angle, α_{rh} (Chapter 10).

Perceptive reviews by Dr. Brian Mason, Smithsonian Institute, Washington, DC, Professor Robert T. Dodd, SUNY, Stony Brook, NY, Professor Roger Burns, Massachusetts Institute of Technology, Cambridge, MA, and Mr. David Elbert, University of Massachusetts, Amherst, as well as the editorial assistance of Peter-John Leone and Michael Gnat, Cambridge University Press, have greatly contributed to the improvement of the book. Thanks are also due the following students for locating errors during trial use of the book manuscript: John Burr, Marshall Chapman, Vincent Dellorusso, Karin Olson, Virginia Peterson, Jennifer Thompson, Suizhou Xue, and Yang Yu.

The author is especially grateful to Mrs. Marie Litterer, University of Massachusetts, for the superb drafting of the many complex illustrations.

1 Atoms, electrons, orbitals, and minerals

A crate of oranges hastily filled at the orchard can be more efficiently packed if vigorously shaken a few times to eliminate waste space. In a similar way, atoms loosely collected or disordered in space can become more energetically stable by bonding together into an ordered crystal structure. Evaluation of the manner in which the various atoms become ordered into stable chemical compounds, both natural and synthetic, and their resultant chemical and physical properties are the principal objectives of the science of crystal chemistry.

For example, atoms of silicon, Si, and oxygen, O, may combine or become ordered into different crystal structures under different conditions of growth. At low pressures and moderate temperatures of formation, each Si atom will surround itself with four O atoms to form the SiO_4 tetrahedron. Each such tetrahedron will then share each of its four corner O atoms with adjoining Si atoms to build a giant three-dimensional polymer or framework of tetrahedra that characterizes the common mineral quartz. Each Si is thus said to be coordinated with four O atoms, and each O atom by two Si atoms to yield the formula $Si^{IV}O_2^{II}$.

At very high pressures and temperatures of formation, such as those induced by an impacting meteorite, six O atoms can be compressed about the Si atom in an octahedral array to form the mineral stishovite. Now, each Si atom is coordinated by six O atoms and each O by three Si atoms to yield the formula $Si^{VI}O_2^{III}$. Quartz, the low pressure, low coordination form of SiO_2, has the more open structure and possesses a density of 2.65 g/cm^3 and a mean index of refraction of 1.547; stishovite, the high pressure, high coordination form of SiO_2, has a density of 4.28 g/cm^3 and a mean index of refraction of 1.815.

Similarly, elemental carbon, C, will, at low pressures of growth, surround itself with three additional C atoms arrayed in an equilateral triangular coordination, with resulting polymerization into sheets of C atoms; whereas during growth at the high pressures that obtain deep in the earth, each C atom will surround itself by four additional C atoms arrayed at the corners of a tetrahedron to build the three-dimensional framework that typifies the mineral diamond. Graphite has the formula C^{III}; diamond, C^{IV}. Graphite has a density of 2.23 and is opaque, reflecting incident light, whereas diamond has a density of 3.51 and a very high index of refraction, 2.418. The causes for such contrasts in physical properties will be further explored and explained in Chapter 2.

The student of crystal chemistry must therefore become familiar with the nuclear and electronic structures and properties of atoms as well as with the organization chart of atoms known as the periodic table or periodic classification of the elements (Fig. 1.1).

Atoms, elements, and nuclides

From the viewpoint of the chemist, the atom may be regarded as a spherical bundle of energy or matter of small mass that retains its identity in chemical combination. The atom consists of *nucleons* (nuclear particles) and extranuclear *electrons*. Nucleons consist primarily of positively charged *protons* and uncharged or neutral *neutrons,* along with subordinate numbers of other particles such as *mesons,* which may have electrical charge of $+$, $-$, or 0. The number of protons is equal to the *atomic number* (Z); and the number of neutrons, to the *neutron number* (N). The sum of protons and neutrons is equal to the *relative atomic mass number,* or *atomic weight* (A); thus $Z + N = A$. The atomic mass is relative because it is generally not given in a quantity such as grams but as a number that uses elemental carbon, $C = 12.000$, as a reference standard. Thus, elemental hydrogen, H, with one proton and no neutrons in its

Figure 1.1. Periodic classification of the elements.

◄—Rrepesentative►		◄—————————Transition————————►										◄——————————— Representative ———————►					
s^1	s^2	d^1	d^2	d^3	d^4	d^5	d^6	d^7	d^8	d^9	d^{10}	p^1	p^2	p^3	p^4	p^5	p^6
H 1 1.00797																H 1 1.00797	He 2 4.0026
Li 3 6.939	Be 4 9.0122											B 5 10.811	C 6 12.01115	N 7 14.0067	O 8 15.9994	F 9 18.9984	Ne 10 20.123
Na 11 22.9898	Mg 12 24.312											Al 13 26.9815	Si 14 28.086	P 15 30.9738	S 16 32.064	Cl 17 35.453	Ar 18 39.948
K 19 39.102	Ca 20 40.08	Sc 21 44.956	Ti 22 47.90	V 23 50.942	Cr 24 51.996	Mn 25 54.9381	Fe 26 55.847	Co 27 58.9332	Ni 28 58.71	Cu 29 63.54	Zn 30 65.37	Ga 31 69.72	Ge 32 72.59	As 33 74.9216	Se 34 78.96	Br 35 79.909	Kr 36 83.80
Rb 37 85.47	Sr 38 87.62	Y 39 88.905	Zr 40 91.22	Nb 41 92.906	Mo 42 95.94	Tc 43 (99)	Ru 44 101.07	Rh 45 102.905	Pd 46 106.4	Ag 47 107.870	Cd 48 112.40	In 49 114.82	Sn 50 118.69	Sb 51 121.75	Te 52 127.60	I 53 126.9044	Xe 54 131.30
Cs 55 132.905	Ba 56 137.34	La* 57 138.91	Hf 72 178.49	Ta 73 180.948	W 74 183.85	Re 75 186.2	Os 76 190.2	Ir 77 192.2	Pt 78 195.09	Au 79 196.967	Hg 80 200.59	Tl 81 204.37	Pb 82 207.19	Bi 83 208.980	Po 84 (210)	At 85 (210)	Rn 86 (222)
Fr 87 (223)	Ra 88 (226)	Ac** 89 (227)															

	Ce 58 140.12	Pr 59 140.907	Nd 60 144.24	Pm 61 (145)	Sm 62 150.35	Eu 63 151.96	Gd 64 157.25	Tb 65 158.924	Dy 66 162.50	Ho 67 164.930	Er 68 167.26	Tm 69 168.934	Yb 70 173.04	Lu 71 174.97
*Lanthanides														
**Actinides	Th 90 232.038	Pa 91 (231)	U 92 238.03	Np 93 (237)	Pu 94 (242)	Am 95 (243)	Cm 96 (247)	Bk 97 (249)	Cf 98 (251)	Es 99 (254)	Fm 100 (253)	Md 101 (256)	No 102 (253)	Lw 103 (257)

simple nucleus, has atomic number 1, neutron number 0, and atomic mass 1.00797 or about one-twelfth of the mass of the reference standard C = 12. From the foregoing discussion, it might be assumed that all of the mass of the atom is contained in the nucleus if $Z + N = A$. This is close to the truth, inasmuch as a proton has about 1837 times the mass of an electron. The following mass numbers are accepted for atomic particles:

$$\text{proton} = 1.00797 \quad \mu \text{ meson} = 0.215$$
$$\text{neutron} = 1.00894 \quad \pi \text{ meson} = 0.280$$
$$\text{electron} = 0.00055$$

In the neutral atom, the number of electrons must equal the number of protons in order to maintain electrical neutrality. Although, at this writing, 106 chemical elements have been identified and periodically classified, there exist more than 1500 variants of these, which are classified as nuclides. A *chemical element* is an atomic species possessing a specific number of protons in its nucleus; a *nuclide* is simply any atomic species that has a specified atomic mass, in which case there is no restriction on the number of either protons or neutrons. The same chemical element Z may exist in species with different numbers of neutrons, and these are classed as *isotopes* (the "p" is to remind you that isotopes contain equal numbers of protons). Table 1.1 lists the isotopes of oxygen in

Table 1.1. *Classification of nuclides as isotopes (= Z), isotones (= N), and isobars (= A)[a]*

Isotopes of oxygen, Z = 8, A = 15.9994[b]
$^{16}_8O_8$ $^{17}_8O_9$ $^{18}_8O_{10}$

Isotones of manganese, iron, and nickel, N = 30
$^{55}_{25}Mn_{30}$ $^{56}_{26}Fe_{30}$ $^{58}_{28}Ni_{30}$

Isobars of argon, potassium, calcium, A = 40
$^{40}_{18}Ar_{22}$ $^{40}_{20}Ca_{20}$ $^{40}_{19}K_{21}$

Isotope	Mass	% Natural abundance	Product
^{16}O	15.9949	99.759	15.956352
^{17}O	16.9914	0.037	0.0062897
^{18}O	17.99916	0.204	0.0367183
		100.00	$A = 15.9993600$

[a] Surrounding the elemental symbol are the atomic mass number (upper left), atomic (proton) number (lower left), and neutron number (lower right). This system is used throughout this book.
[b] Oxygen has atomic mass $A = 15.9994$, based on the mass of carbon (C = 12.000).

which atomic nuclei all contain eight protons ($Z = 8$) combined with either eight, nine, or ten neutrons ($N = 8$, 9, or 10) to yield the oxygen isotopes ^{16}O, ^{17}O, and ^{18}O.

Reversing the situation, we find that other elements or nuclides may have equal numbers of neutrons combined with unequal numbers of protons, and these are classed as *isotones* (the "n" is to remind you that these contain equal numbers of neutrons). Thus Table 1.1 lists the isotones ^{55}Mn, ^{56}Fe, and ^{58}Ni, all containing thirty neutrons. There also exist nuclides that contain unequal numbers of both protons and neutrons but have equal integral mass number: these are classed as *isobars* (the "a" is to remind you that these have equal atomic mass numbers). In Table 1.1, isobars are represented by ^{40}Ar, ^{40}K, and ^{40}Ca.

Nuclear structure and stability

Of the more than 1500 nuclides, both natural and synthetic, that have been identified, only 269 are stable, that is, do not decay by radioactive processes. Although all 106 of the chemical elements contain a percentage of radioactive nuclides, those with atomic numbers $Z = 1$–82 (H through Pb) contain only a small percentage of these. Those elements with atomic numbers $Z = 83$–106 consist entirely of radioactive nuclides; or, to put it differently, those elements with $Z > 82$ are all unstable and decay naturally by a variety of radioactive processes. Thus, our 269 stable nuclides must come from elements with $Z = 1$–82. It is common to classify the stable nuclides into four groups characterized by the even and odd proton and neutron constitution of their nuclei. It is obvious from Table 1.2 that nuclides having even numbers of both protons and neutrons have exceptionally high stability and abundance, whereas those having an odd–odd combination are rare in nature. The other two groupings having even–odd and odd–even combinations are of intermediate abundance. This tells us that the atomic nucleus also has a structure, and that protons and neutrons are not merely an irregular or disordered aggregate of

Table 1.2. *Classification of stable nuclides on the basis of proton and neutron configuration*

Nuclear content		No. of nuclides	Example
Protons (Z)	Neutrons (N)		
Even	–Even (e–e)	160	$^{16}_{8}O_8$
Even	–Odd (e–o)	55	$^{9}_{4}Be_5$
Odd	–Even (o–e)	49	$^{27}_{13}Al_{14}$
Odd	–Odd (o–o)	5	$^{14}_{7}N_7$
		269	

both. We need not look too far for the reasons for this, when we note that inside the nucleus repulsive proton–proton forces far exceed attractive forces. At short range, protons and neutrons are strongly interactive, and their nuclear pairing ($n + p$) creates cohesive forces that greatly exceed magnetic, electric, or gravitative forces. Here, the aforementioned mesons also play an important role. As noted earlier, μ mesons and π mesons are of different mass and may also have electric charge +, −, or 0. These mesons appear to help overcome the electrostatic proton–proton repulsion in the nucleus and further contribute to nuclide stability.

Where atomic nuclei become either oversaturated or undersaturated with protons or neutrons, they tend to expel the excess particles by means of natural radioactive processes. Where neutrons are in excess, the number of protons is increased by a *β-decay process,* whereby a neutron decays to a proton and ejects an electron; this process is written

$$n \longrightarrow p + e^- \text{ ejection}$$

(e.g., $^{87}_{37}Rb_{50} \rightarrow {}^{87}_{38}Sr_{49} + $ energy). If protons are in excess, the reverse occurs, and we have an *electron capture process:*

$$p \longrightarrow n + e^+ \quad \text{positron ejection}$$

(e.g., $^{40}_{19}K_{21} \rightarrow {}^{40}_{18}Ar_{22} + $ energy). ^{87}Rb and ^{40}K constitute small percentages of radioactive isotopes associated with their abundant stable family members ^{86}Rb and ^{39}K. Radioactive ^{40}K and ^{87}Rb contribute much of the heat in the earth's crust, inasmuch as they are concentrated in potassium feldspar, sodium–calcium feldspar, and micas, all principal minerals of the more common rocks that make up the earth's crust. Over the 4.6×10^9-yr history of the solar system these radioactive nuclides, along with those of uranium, U, and thorium, Th, have been producing large, but decreasing, amounts of heat at known rates of atomic disintegration. Each radioactive nuclide has a known *half-life* or time interval in which one-half of the atoms present decay. Radioactive decay also plays an important role in crystal chemistry directly, particularly in those cases where decay of unstable nuclides, such as ^{238}U, ^{235}U, and ^{232}Th produce copious quantities of alpha particles that bombard and gradually destroy the crystal structures of the minerals in which they reside. Minerals whose structures are so damaged or destroyed are classed as *metamict minerals.* They are discussed further in a subsequent chapter dealing with radioactive minerals.

At this point, the important conclusion may be made that nuclear structure of atoms controls the abundance of the chemical elements in the universe, whereas electronic structure of atoms controls the species of minerals that are permitted to grow or crystallize. Protons and neutrons control nuclear stability, but electrons control crystal stability via processes of chemical bonding. Accordingly, a more detailed examination of the electronic structure of atoms is essential to our comprehension of modern concepts of crystal chemistry.

Electronic structure

Matter has both corpuscular and undulatory properties, and it is possible to consider the electron either as a particle or as a wave. An *electron* is perhaps best described as a continuous distribution of negative electrical charge carried by waves. In these waves, *electron density* ψ^2 is proportional to the *intensity* of the waves. The point that represents the *energy* of the electron is guided by the waves that surround it. In the early treatment of the atom, attributed to Niels Bohr, the electron was perceived as a negative particle revolving around the nucleus in a three-dimensional racetrack or solar system. To a first approximation, the comparison between the atomic nucleus and its electron cloud with the sun and the solar system is impressive. Table 1.3 illustrates the remarkable correlation between the ratio of the diameter and mass of the atomic nucleus and its electron cloud and that of the sun and its planetary solar system, and serves to emphasize the concentration of mass in the atomic nucleus and the vast amount of space between the nucleus and the outer limits of the electron cloud.

The analogy falters, however, when we note that the elliptical orbits of the planets about the sun all lie more or less in one plane. For the atom, this is contrary to the concepts of modern quantum physics and the wave mechanical treatment of the electron developed by Schrödinger and his associates fairly early in this century. In the Schrödinger atom, the electron is assumed to be smeared out in an electron cloud or orbital of specific shape or dimension, denser in some regions than in others. Wave mechanics actually predicts the *probability of distribution of electron density* of different orbitals or energy levels ascribed to particular quantum states. It tells us that, at any given instant in time, a wave has, at a given position in space, a specific amplitude. This amplitude, at every position in space, at all times, completely describes the wave. Surfaces are drawn enclosing the amplitude of the wave function, which outline the electron cloud, or *atomic orbital*. Once again, these surfaces or orbitals indicate the restricted areas where, with high probability, the electron will be found. Thus the electron may be considered as a particle concentrated in

Table 1.3. *Comparison of diameters and mass percentage of an average atomic nucleus with its electron cloud and of the sun with the solar system*

Object	Diameter	Mass %
Atomic nucleus	10^{-12} cm	99.95
Electron cloud	10^{-8} cm	0.05
Sun	10^9 m	99.85
Solar system	10^{13} m	0.15

the denser regions of the electron cloud or as a wave of negative charge smeared out over the confines of the electron cloud or orbital.

In some cases, the one treatment, and in some cases, the other, provides better insight for the explanation of various atomic phenomena. We may note that ψ is the orbital wave function in space and time, ψ^2 is the probability of finding the electron at a particular point, and $\psi^2 R^2$ is the probability of finding the electron in a shell of a given thickness and radius; $4\pi R^2 \psi^2$ describes the number of electrons in a spherical shell of a given thickness at a radius R from the nucleus.

To solve for the wave function and determine the probable location of the electron in its orbital, Schrödinger showed that four quantum numbers or parameters had to be determined: the *principal quantum number n*, which establishes the radial distance of the electron from the nucleus; the *azimuthal* or *angular momentum quantum number l*, which gives the angular momentum of the electron in orbit and helps define the shape of the orbital; the *magnetic quantum number m_l*, which determines the orientation of the angular momentum in space; and the *spin quantum number s*, which determines the sense of rotation of the electron or "spin up or spin down," in the magnetic field of the spinning electron.

The principal quantum number n has integer values and determines the position of the horizontal rows of the conventional periodic table (Fig. 1.1). The azimuthal quantum momentum number l may take the values 0, 1, 2, 3, 4, which are equivalent to the letter designations $s, p, d, f, g, \ldots$. An s orbital has

Figure 1.2. Orbitals and quantum numbers.

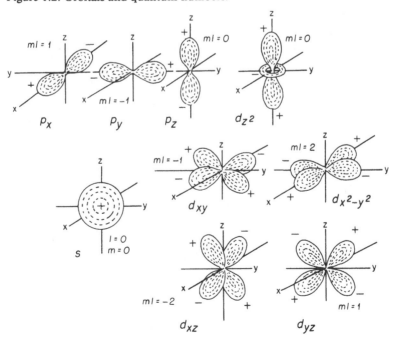

no (i.e., 0) angular momentum, because it is spherical; a p orbital is quasi-dumbbell-shaped and axially directed; a d orbital has various shapes, three possible shapes being double dumbbells of electron density, axially directed, a fourth, axially directed, and a fifth, axially directed but quasi-dumbbell-shaped with a central doughnut or node of electron density (Fig. 1.2).

The magnetic quantum number m_l restricts the orientation and number of each type of orbital, so there are one s orbital, three p orbitals, five d orbitals, and seven f orbitals in a particular quantum shell. Each of these orbitals may be occupied by a maximum of two electrons, provided that their spins are opposed. Each orbital level, s, p, d, and f, may be regarded as an energy level when given in combination with the principal quantum number (Fig. 1.3). Electrons introduced into an atom will occupy the lowest energy levels first in the sequence $1s$, $2s$, $2p$, $3s$, $3p$, $4s$, but then $3d$, before $4p$, and so on. No two electrons in the same atom can be described by the same wave function ψ unless their spins are opposed. Further, in filling orbitals, each quantum state must be occupied singly before it can be occupied doubly; thus where five d orbitals are being filled, we must place one electron in each before a second electron with opposite spin can be introduced. These so-called selection rules (Hund's rules)

Figure 1.3. Orbitals and energy levels.

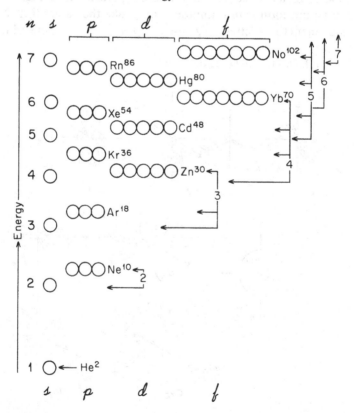

just alluded to express the tendency for electrons to remain as far apart as is possible. We emphasize that we are now considering the *unexcited, neutral,* or *ground state* atoms.

Periodic classification

We are now sufficiently well informed to analyze and use the periodic table or periodic classification of the elements, for which several formats exist. We will make use of both the standard row and column, or so-called long form, format (Fig. 1.1) as well as the less familiar but more elegant spiral format (see Fig. 1.4), introduced by the American Chemical Society. After some study, it will become apparent that the periodic classification attributed to Mendeleev (for whom element $Z = 101$ is named mendelevium, Md) organizes the elements according to their quantum numbers, and, in so doing, also organizes them into vertical groups according to similarities in their chemical properties. This leads to an arrangement with seven horizontal rows or periods representing the principal quantum numbers, $n = 1, \ldots , 7$, also designated as the *shells* K–Q. Within each row or shell, the elements are arranged in order of increasing atomic number, Z. The vertical columns are called *groups* or *subshells* and contain the orbitals s, p, d, and f, giving one, three, five, and seven sites, respectively, in which we can place two, six, ten, and fourteen electrons. Ignoring, for the moment, the f orbitals, we can divide our table into eighteen vertical *groups* by successively adding $2 + 6 + 10 = 18$ single and opposed spin pairs of electrons into the s, p, and d orbitals. The seven f orbitals, in which can be placed fourteen electrons building up fourteen different elements, are grouped separately at the bottom of the long form table. Before going further, we note that spectroscopists and chemists use different nomenclature for the periodic classification of the elements. From Table 1.4 it can be seen that there is overlap in the energies of elements of the third and fourth principal quantum levels, and the $4s$ level is lower in energy than the $3d$ in the neutral atom. Thus the krypton shell of the chemist, the fourth horizontal row of the table, contains or includes the third level of the spectroscopists' M shell. The important feature to note is that filling of the fourth principal quantum level that begins with potassium (K, $Z = 19$) proceeds in accordance with increasing energy levels in the sequence $4s$, $3d$, $4p$. Sequential filling of the eighteen electron sites, two electrons in s orbitals, ten in d orbitals, and six in p orbitals, will build the neutral atom structure of the eighteen elements following K (see Table 1.5).

By placing one electron in the lowest energy level, the spherical $1s$ orbital level, we form hydrogen, coded $_1$H $1s^1$, sequentially giving the atomic number, the element symbol, the orbital energy level, and the number of electrons contained. A second d electron with opposed spin placed in the $1s$ orbital completes the K (or He) shell, forming the gas helium $_2$He $1s^2$. The next electron is placed in the $2s$ orbital, a sphere of electron density with a radius larger than that of $1s$, and this gives $_3$Li $1s^22s^1$ for the element lithium. A fourth electron yields beryllium, $_4$Be $1s^22s^2$, completing the $2s$ subshell.

The fifth electron is placed in a dumbbell-shaped p orbital to form boron, $_5$B $1s^22s^22p^1$. The sixth electron must be placed in a different p orbital (single before double occupancy) in forming carbon, $_6$C $1s^22s^22p^2$. The seventh electron is placed in the third p orbital, giving nitrogen, $_7$N $1s^22s^22p^3$ with one electron in *each* of the three $2p$ orbitals. The eighth electron is, with spin opposed, placed in the first p orbital to form oxygen, $_8$O $1s^22s^22p^4$; the ninth electron forms fluorine $_9$F $1s^22s^22p^5$, and a tenth electron closes out the L (or Ne) shell with the rare gas neon, $_{10}$Ne $1s^22s^22p^6$.

For each periodic row after the first, $n = 1$, an eight-electron complement of s^2p^6 completes the energy level with the formation of a *rare* or *noble* gas. Several of the elements have been given special group names because their similarity in electronic structure leads to similarities in chemical behavior. In addition to the rare or noble gases of Group VIII (i.e., those with a full complement of s^2p^6 outer electrons), we have the *alkali* elements or *alkalies* of Group IA, which are those with an outer electron configuration of s^1; these are followed by the *alkaline earth* elements of Group IIA, which are those with an outer electron configuration of s^2 and the *halogens* of Group VIIB (alternatively labeled VIIA by other schemes), which consist of those elements having an outer electron configuration of s^2p^5. Those sequences of elements containing outer electrons in d orbitals, d^1–d^{10}, form horizontal sequences classed as *transition elements* or *transition metals*. Two other sequences, represented by those elements having outer electrons in f orbitals, f^1–f^{14}, are classed as *lanthanides* or *rare earth elements* and *actinides*. Here, the elegance of the arrangement of the spiral

Table 1.4. *Classification of electron shells*

Spectroscopists' nomenclature			Chemists' nomenclature		
Shell	Electron complement	No. of electrons	Shell	Electron complement	No. of electrons
K	$1s^2$	2	He	$1s^2$	2
L	$2s^22p^6$	8	Ne	$2s^22p^6$	8
M	$3s^23p^63d^{10}$	18	Ar	$3s^23p^6$	8
N	$4s^24p^64d^{10}4f^{14}$	32	Kr	$3d^{10}4s^24p^6$	18
O	$5s^25p^65d^{10}5f^{14}(5g^{18})^a$	50	Xe	$4d^{10}5s^25p^6$	18
P	$6s^26p^66d^{10} \ldots^b$	72	Rn	$4f^{14}5d^{10}6s^26p^6$	32
Q	$7s^27p^6 \ldots$		Eka-Rn^c	$5f^{14}6d^{10}7s^27p^6$	32

[a] Nine g orbitals with eighteen electrons are theoretically possible.
[b] Eleven h orbitals with twenty-two electrons are theoretically possible.
[c] The rare gas with atomic number 118 has not been discovered, but chemists have provisionally assigned it the shell name of Ekaradon.

periodic table (Fig. 1.4) beautifully illustrates these sequences and the manner in which transition-metal elements, lanthanides, and actinides interrupt the alkali s^1 to rare gas p^6 sequences shown as the quasi-circular portion of the table. The spiral table also serves to emphasize that the simplest element, H, is the nucleus out of which all of the other elements are formed. Thus, the fusion of four hydrogen atoms of atomic mass $1.0079 \times 4 = 4.0316$ yields He, with atomic mass 4.0026. The small difference in masses, 0.0290, represents the energy released, which is enormous when millions of atoms participate. This process is called *hydrogen burning,* which, in addition to forming He and, ultimately, the other elements, also supplies the energy to keep the sun both hot and luminous.

We return now to the filling sequence of the periodic table. After the completion of the second period with neon, Ne, with its complement of $1s^2 2s^2 2p^6$ electrons, a third period, the M shell, is opened with a third spherical $3s$ orbital located at a considerable increase in radial distance from the closed K and L shells. Here, successive s orbitals or clouds of high electron density are separated

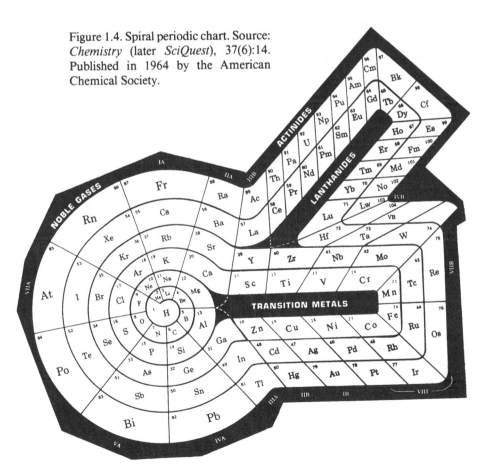

Figure 1.4. Spiral periodic chart. Source: *Chemistry* (later *SciQuest*), 37(6):14. Published in 1964 by the American Chemical Society.

from one another by *nodes* or regions in which the electron density is zero. Filling of the third periodic row leads to the rare gas argon, $_{18}$Ar $1s^22s^22p^63s^23p^6$, which we can further abbreviate as $_{18}$Ar(Ne)$3s^23p^6$, stating that argon has a ten-electron core of neon inside the outermost $3s^23p^6$ shell. Our new, shortened code then gives, in parentheses, the symbol of the nearest previously filled rare gas core followed by the outer electrons that succeed it. We will now fill the entire fourth period of the table by adding outer electrons to the Ar core, going from elements $Z = 19$ to $Z = 36$, or K to Kr. The orbital filling sequence is represented schematically in Table 1.5.

Elements of the fifth period, rubidium, Rb, to xenon, Xe, are filled in the same sequence as the fourth row, but they now begin with a $5s$ orbital and closing with a $5p$ orbital of six electrons, with a ten-electron $4d$ sequence of transition elements lying between s and p orbitals. After xenon, a sixth period is opened with the formation of cesium, Cs, with $6s^1$, followed by barium, Ba, $6s^2$, and then lanthanum, La, $6s^25d^1$.

Now the $5d$ filling sequence is interrupted by the introduction of the seven $4f$ orbitals. When these are filled with their fourteen electrons, they give rise to the lanthanides or rare earth elements, from cerium, Ce, to lutetium, Lu. Note that although this completes the spectroscopists' N shell, it lies within the chemists' radon shell. The latter is completed with resumption of filling of the $6p$ orbitals with six electrons to close the shell with the rare gas radon, Rn. Now the pattern of the sixth period is begun anew. The seventh period opens with a $7s$ orbital

Table 1.5. *Sequence of filling of orbitals from $_{19}K$ to $_{36}Kr$*

Element		Z	4s	3d	4p	Electron code
Potassium	K	19	′			$(Ar)4s^1$
Calcium	Ca	20	″			$(Ar)4s^2$
Scandium	Sc	21	″	′		$(Ar)4s^23d^1$
Titanium	Ti	22	″	′ ′		$(Ar)4s^23d^2$
Vanadium	V	23	″	′ ′ ′		$(Ar)4s^23d^3$
Chromium	Cr	24	′	′ ′ ′ ′ ′		$(Ar)4s^13d^5$
Manganese	Mn	25	″	′ ′ ′ ′ ′		$(Ar)4s^23d^5$
Iron	Fe	26	″	″ ′ ′ ′ ′		$(Ar)4s^23d^6$
Cobalt	Co	27	″	″ ″ ′ ′ ′		$(Ar)4s^23d^7$
Nickel	Ni	28	″	″ ″ ″ ′ ′		$(Ar)4s^23d^8$
Copper	Cu	29	′	″ ″ ″ ″ ″		$(Ar)4s^13d^{10}$
Zinc	Zn	30	″	″ ″ ″ ″ ″		$(Ar)4s^23d^{10}$
Gallium	Ga	31	″	″ ″ ″ ″ ″	′	$(Ar)4s^23d^{10}4p^1$
Germanium	Ge	32	″	″ ″ ″ ″ ″	′ ′	$(Ar)4s^23d^{10}4p^2$
Arsenic	As	33	″	″ ″ ″ ″ ″	′ ′ ′	$(Ar)4s^23d^{10}4p^3$
Selenium	Se	34	″	″ ″ ″ ″ ″	″ ′ ′	$(Ar)4s^23d^{10}4p^4$
Bromine	Br	35	″	″ ″ ″ ″ ″	″ ″ ′	$(Ar)4s^23d^{10}4p^5$
Krypton	Kr	36	″	″ ″ ″ ″ ″	″ ″ ″	$(Ar)4s^23d^{10}4p^6$

and the formation of the rare elements francium, Fr, radium, Ra, and actinium, Ac, with orbitals containing $7s^1$, $7s^2$, and $6d^1$ electrons.

Filling of the $6d$ subshell is now interrupted by the appearance of $5f$ orbitals, and fourteen new elements, the actinides, are formed. These contain the naturally radioactive Th and U, as well as the rare radioactive elements from neptunium, Np, through lawrencium, Lw ($Z = 103$), largely spontaneous fission products. Elements 104–106 have been discovered by different investigators in the United States and the USSR, and the names proposed for these remain in dispute. Elements 107–118 have not yet been discovered, and the seventh period remains incomplete, although chemists have long ago proposed the name "Ekaradon" for this shell.

This completes our discussion of the nuclear and electronic structure of the neutral atoms and their arrangement in a periodic classification. Although periodic irregularities do exist, periodic classification of the elements constitutes a great creative scientific achievement. In Table 1.5, a periodic irregularity may be noted in the *absence* of a $3d^4$ configuration in going from $_{23}$V, $3d^3$ to $_{24}$Cr, $3d^5$. Note also the absence of $3d^9$ in going from $_{28}$Ni, $3d^8$ to $_{29}$Cu, $3d^{10}$. In both of these irregularities the Cr and Cu atoms carry one rather than two electrons in their $4s$ orbitals. Continued study of and regular reference to periodic classification offers great rewards.

Summary

Crystal chemistry evaluates the influence of the packing of atoms in minerals on their physical and chemical properties.

The atom is the smallest amount of matter that retains its identity in chemical combination. It consists of a nucleus of protons, neutrons, and mesons, which contains the greater part of its mass and of a cloud of electrons surrounding the nucleus, whose behavior is governed by four parameters or quantum numbers.

There are 106 chemical elements, based on the number of protons in the nucleus. Of these, elements 1–82 are relatively stable and 83–106 naturally radioactive.

The stable elements form 269 nuclides (atomic species with specified atomic mass). These may be isotopes (equal protons, differing neutrons), isotones (differing protons, equal neutrons), or isobars (differing protons and neutrons, equal atomic mass).

The 269 stable nuclides may have even numbers of protons and neutrons: these are the most abundant. Nuclides with odd numbers of protons and neutrons are rarest in nature. Nuclides with odd–even or even–odd numbers of protons and neutrons are intermediate in abundance.

Atomic nuclei having excess or lack of protons and neutrons are unstable and decay radioactively, thus producing heat.

The number of electrons must equal the number of protons in a neutral
or ground state atom.

Electrons are distributed in shells with different energy levels and
shapes of orbitals according to their quantum numbers.

Electrons govern the chemical behavior of atoms, and how they may be
packed into minerals.

Nuclear structures determine the abundance of the elements.

Elements may be arranged into a periodic table according to their
number of protons and the structure of their electron cloud. In this
table, affinities between families of elements are evident.

Bibliography

Benfy, T., ed. (1964). Spiral periodic chart. In *Chemistry. Amer. Chem. Soc.*, Washington, D.C.

Cotton, F. A., and Wilkinson, G. (1966). *Advanced inorganic chemistry,* 2d ed. Interscience, New York.

Coulson, C. A. (1961). *Valence,* 2d ed. Oxford Univ. Press, London.

Dickerson, R. E., Gray, H. B., and Haight, G. P., Jr. (1974). *Chemical principles,* 2d ed. W. A. Benjamin, Menlo Park, Calif.

Gray, H. B. (1965). *Electrons and chemical bonding.* W. A. Benjamin, Menlo Park, Calif.

Greenwood, N. N., and Earnshaw, A. (1984). *Chemistry of the elements.* Pergamon Press, New York.

Krebs, H. (1968). *Grundzüge der anorganischen Kristallchemie.* Ferdinand Enke Verlag, Stuttgart.

Mason, Brian, and Moore, C. B. (1982). *Principles of geochemistry,* 4th ed. Wiley, New York.

Nassau, Kurt (1983). *The physics and chemistry of color.* New York.

Periodic Table of the Elements (1979). Sargent-Welch Scientific Co., Skokie, Ill.

Pauling, L. (1960). *The nature of the chemical bond,* 3d ed. Cornell Univ. Press, Ithaca, N.Y., pp. 28–63.

Rankama, K., and Sahama, Th. G. (1950). *Geochemistry.* Univ. of Chicago Press, Chicago, Ill.

Sanderson, R. T. (1960). *Chemical periodicity.* Reinhold, New York.

Sebera, D. K. (1964). *Electronic structure and chemical bonding.* Blaisdell, Waltham, Mass.

2 The excited atom: spectra and quanta, ionization potential, electronegativity, and chemical bonding

In the undisturbed, neutral, or ground state of an atom, all electrons will occupy the lowest possible energy levels in the sequence $n = 1, 2, 3, 4, \ldots$ and $l = 0, 1, 2, 3, \ldots$, equivalent to $s, p, d, f, \ldots$. Electrons can change their energy level only by "jumping" from one quantum state to another under the influence of some external added energy source. When sufficient energy is thus absorbed, an electron may leave its s orbital and "leap" to a higher vacant p orbital site; this energy will be released when the excited electron (or a neighboring electron in the p site) drops down or returns to its original s orbital. This release of energy, often radiant visible energy, is detectable as emission spectral lines of specific wavelength and frequency. With a spectroscope or a spectrograph, it is possible to observe and measure the wavelength or frequency of each spectral line and to determine the specific quantum states or orbital levels involved in a given electronic transition. Thus, careful study of emission spectral lines has provided independent verification of the quantum theory of atomic and electronic structure, in that each line can be catalogued in terms of principal, azimuthal or angular, and spin quantum numbers. An omnipresent serendipity has also given the crystal chemist a very valuable method for the chemical analysis of minerals, rocks, and other inorganic materials.

Atoms can be excited by various means, including radiation or heat from an ordinary light bulb, an ultraviolet lamp, a candle, a Bunsen flame, or a higher energy electric carbon arc. Atoms can also be excited by the close approach of, and/or collision with, other atoms, which may give rise to chemical bonding of atoms and formation of crystals. Some atoms are very difficult to excite and are thus reluctant to emit spectral lines or combine with other atoms. This is particularly true of the rare or noble gas atoms with electronic configuration $1s^2$ and $2-7s^2p^6$, He, Ne, Ar, Kr, Xe, and Rn (see Fig. 1.1). Completed s^2p^6 orbitals endow the rare gases with high stability, allowing them to remain as gases. Conversely, elements of the alkali group, $n \ldots s^1$, Li, Na, K, Rb, and Cs, all have a single electron outside of a core of a rare gas element. This electron will be very loosely bound to the atomic nucleus, and can be excited to leave its orbital with only a small input of external energy. Such a loosely bound electron contained in an orbital lying outside of a rare gas core is called a *valence electron*. Moving to different groups in our periodic tables (see Fig. 1.1 or 1.4), we can observe an increase in the number of valence electrons with the vertical group number (e.g., K $4s^1$, Ca $4s^2$, Sc $4s^23d^1$, Ti $4s^23d^2$, V $4s^23d^3$, . . .). On reaching the element zinc, Zn, $4s^23d^{10}$, the completed $3d$ subshell binds these electrons securely so that Zn effectively has only the two s electrons that are valence electrons. Spectroscopy has shown that electrons in s orbitals are easy to excite, and yield spectral lines of high sensitivity, permitting their detection even when present in very minute quantities. Elements with p orbital valence electrons are very difficult to excite and have low spectral sensitivity, and those with d and f orbital valence electrons are of intermediate excitation and detection level.

When common salt, NaCl (halite), is held in a candle or a Bunsen flame, the small thermal energy source excites the $3s^1$ electron to jump or ascend to an

Table 2.1. *Correlation of the periodic variation of spectral line multiplicity with maximum valence of atoms*

4s, 3d electrons	s^1	s^2	s^2d^1	s^2d^2	s^2d^3
Princ. quantum #4	K	Ca	Sc	Ti	V
Valence electrons	1	2	3	4	5
Electron spin, s	$\pm\frac{1}{2}$	±1	±2	±3	±4
Sum of spins, S	$\frac{1}{2}$	$0,1$	$\frac{1}{2},1\frac{1}{2}$	$0,1,2$	$\frac{1}{2},1\frac{1}{2},2\frac{1}{2}$
Multiplicity M = 2S + 1	2	1,3	2,4	1,3,5	2,4,6
Spectral region	Red	Green	Green	Green	Blue
Maximum multiplicity pattern	‖	⦀	⦀⦀	⦀⦀	⦀⦀⦀
Wavelength in Å	7699	5270	5087	5014	4481
	7664	5265	5085	5007	4876
		5261	5083	4999	4865
			5081	4991	4851
				4981	4832
					4827

empty $3p$ orbital: when this electron falls back to home base, a pair of spectral lines is emitted, constituting the sodium, Na, doublet at wavelengths $\lambda = $ 5890 Å and 5896 Å (Jaffe, 1949). Although the separation is only 6 Å (1 Å $ = 10^{-8}$ cm), this line pair is easily resolved with the simplest of spectroscopes. The wavelengths cited lie in the yellow region of the visible spectrum of electromagnetic radiation and color the flame a bright yellow. Because many Na atoms will be excited, some may be induced to leap to levels higher than $3p$ and may return via a longer route in a manner that will cause additional Na doublets to appear in different parts of the spectral region: the latter will be less sensitive to spectral detection.

The way in which spectra have been used to verify quantum theory is best illustrated by the multiplicity M of spectral lines, caused by the clockwise and counterclockwise spin of electrons. Spectroscopists discovered that the simple formula relating multiplicity to electron spin S is $2S + 1 = M$, where S is the sum of the possible spins, s, and M is the multiplicity of lines emitted. Spin quantum numbers may have a value of $+\frac{1}{2}$ or $-\frac{1}{2}$ (or $\pm\frac{1}{2}$) per electron. Thus, an alkali element with one valence electron, s^1, can have the electron spinning either way, giving $s = \pm\frac{1}{2}$. Accordingly, S can only equal $\frac{1}{2}$ and $2S + 1 = 2$, indicating that alkali elements should emit their spectral lines in pairs or doublets, even though only one electron is involved per atom. If we move across a row of the periodic table, using principal quantum number 4 as an example, we see that spectral multiplicity, $2S + 1 = M$, will consistently be one number greater than the number of valence electrons for the elements K, Ca, Sc, Ti, and V (Table 2.1).

It has been shown that the multiplicity term $2S + 1 = M$ leads to an emission spectral pattern of singlets and triplets for the divalent calcium, Ca, atom (Table 2.1 and Fig. 2.1). However, at the short-wavelength end of the visible region, we note the appearance of a doublet pattern (3968.5 Å and 3933.7 Å) emitted by Ca. Inasmuch as doublet spectral patterns or multiplicities imply the presence

Figure 2.1. Atomic (Ca), ionic (Ca$^+$), and molecular (CaO) spectra of calcium.

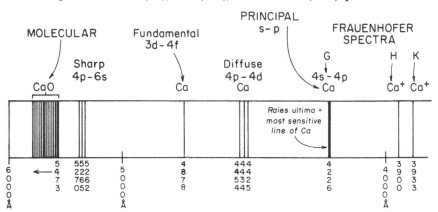

of an atom with one valence electron, $S = \pm\frac{1}{2}$, one must conclude that such a pattern is emitted not by the neutral Ca atom but by the singly ionized Ca^+ ion.

When enough energy is supplied (e.g., higher temperature or higher voltage), Ca will give up one of its $4s^2$ valence electrons to become Ca^+, with a $4s^1$ electron structure, and this univalent ion then behaves like a univalent alkali element and emits doublet spectra. Because of the higher energy absorbed, the spectral lines are shifted to shorter wavelengths, causing the Ca^+ doublet to appear in the violet region of the visible spectrum. The appearance of the Ca^+ ion doublet in stellar spectra has been used to measure stellar temperatures. Low temperature red stars contain only Ca atom spectra (singlets and triplets) and CaO band and molecular spectra; higher temperature stars, such as the sun, show weak lines of the Ca^+ ion doublet; and high temperature blue stars emit intense doublet spectra of Ca^+.

The energy required to remove an s electron from the neutral Ca atom to infinity is termed the *ionization potential I* (Table 2.2). With still higher energy input, additional electrons may be removed at energy increments labeled I_2, I_3, and so on, referred to as the first, second, third, and so on, ionization potentials for a given chemical element. It is generally the first ionization potential, I_1, that is most important in determining whether an element will ionize. The ease

Table 2.2. *Some crystal chemical concepts*

Ionization potential (I or I_P): Energy or potential required to move an electron from its normal quantum level to infinity. Energy is acquired as $X \rightarrow X^+ + e^-$. Also a measure of electronegativity, power of a free cation to attract anions, nuclear screening, and polarizing power. I_P is usually given in electron volts (ev): 1 ev = 23 kcal/mole.

Electron affinity (E or E_A): Energy released in adding an electron from ∞ to a neutral atom, $X + e^- \rightarrow X^-$. Usually expressed in ev or kcal.

Electronegativity (χ): Power of an atom in a molecule to attract electrons. $\chi = (I_P + E_A)/125$ in kcal/mole. I_P refers to the first ionization potential. $\Delta\chi$ of two atoms in a molecule is directly proportional to the percentage of ionic bonding and inversely proportional to the percentage of covalent bonding. A pure ionic bond is impossible: a pure covalent bond is possible. Why? (See Chapter 4.)

Polarizing power (ze/r^2): Charge divided by radius squared is equal to electric field strength. Small, highly charged cations can readily deform or polarize large anions.

Polarizability: Dipole moment induced in an ion by a unit electrical field. A measure of the deformability of an ion. Large ions are readily deformed; most cations are not.

Ionic potential (ϕ) = (ze/r): Charge divided by radius of ion. A measure of electrostatic charge on the surface of an ion. High potential (ϕ) results in a tendency to form complex anions.

Isoelectronic: Atoms or ions having the *same* electronic or extranuclear structure are said to form an isoelectronic series. Cl^-, neutral Ar, K^+, Ca^{2+}, Sc^{3+}, Ti^{4+}, V^{5+}, Cr^{6+}, and Mn^{7+} are isoelectronic because all have the structure $1s^2$-$2s^2$-$2p^6$-$3s^2$-$3p^6$, that of argon in the ground state.

---

or difficulty with which an element can be made to give up a valence electron is, once again, a periodic function.

Here, a few observations are important: an element having its valence electrons in s orbitals, which have low ionization potential, will ionize readily. Such elements include the alkalies and alkaline earth elements of Groups IA (s^1) and IIA (s^2) of each principal quantum row ($n = 1, \ldots, 7$). Thus, the alkali elements, Li, Na, K, Rb, and Cs, will have ionization potentials lower than those of any other element, because all have a single valence electron in an s orbital. Accordingly, these alkali elements will ionize readily to form *ionic bonds* in crystal growth.

Alkaline earth elements of Group IIA, Be, Mg, Ca, Sr, and Ba, have valence electrons in s orbitals and slightly higher ionization potentials; they will each lose both s^2 electrons to form divalent or $2+$ ions, but will ionize less readily than the alkali elements. The reason for the higher energy of ionization is that the two electrons with opposed spin in s^2 orbitals are paired and fill the s^2 subshell. Removal of an electron from a completed shell or subshell always requires higher energy, indicating that completed electron shells represent a condition of greater stability than do uncompleted shells.

As we move across the periodic table to higher group numbers having s^2d^n valence electrons, the ionization potentials increase, informing us that d electrons are more firmly bound than are s electrons. When a d^{10} shell is completed, as for example at atomic number 30, Zn has electronic structure $4s^2 3d^{10}$, and at this point the ionization potential takes a rapid jump to a higher value caused by the completion of the $3d$ or M shell. Data in Table 2.3 indicate that for Zn the values for I_1, I_2, and I_3 are, respectively, 9.392, 17.960, and 39.701 ev (electron volts). The high value for I_3 indicates why Zn never achieves a valency higher than 2 in the excited atom. Here, the reader might assume that two d electrons leave first, because they arrived last, but it is the two s electrons that are removed, because the completed $3d^{10}$ shell will now have a higher orbital energy. After this, any additional electrons must be placed in the $4p$ orbitals, first singly and then doubly, and I_1 potentials will increase more or less regularly from gallium, Ga, $4p^1$ to krypton, Kr, $4p^6$. Electrons in p orbitals are the most difficult of all to remove, and the rare gas Kr, $4p^6$, has potential $I_1 = 13.996$ ev. All of the rare or noble gases (He, $1s^2$; Ne, $2s^2p^6$; Ar, $3s^2 3p^6$; Kr, $4s^2 4p^6$; Xe, $5s^2 5p^6$; and Rn, $6s^2 6p^6$) have relatively high I_1 values and tend to avoid transferring or sharing their valence electrons with other elements; they remain as gases.

In summary, we see that ionization energies increase and spectral detection sensitivities decrease in the periodic group sequence s, d or f, and p, that such energies increase rapidly and suddenly where shells or subshells become closed. Careful study of Table 2.3 will show how ionization energy requirements restrict the number of valence electrons used by the various elements in chemical bonding.

Leaving ionization potentials for the moment, and reversing our trek across the periodic table, we note that Kr, $4p^6$, has no place to add another electron, whereas bromine, Br, $4p^5$, would readily accept a sixth p electron to become *isoelectronic* with Kr and, in so doing, increase its stability. Selenium, Se, $4p^4$, has room for a pair of p electrons.

Just as energy is absorbed when atoms lose an electron, energy is released when atoms gain an electron. This energy is the *electron affinity* (E_A, see Table

Table 2.3. *First five ionization potentials of the elements (in electron volts)*

Z	Element	I	II	III	IV	V
1.	H	13·595	—	—	—	—
2.	He	24·581	54·403	—	—	—
3.	Li	5·390	75·6193	122·420	—	—
4.	Be	9·320	18·206	153·850	217·657	—
5.	B	8·296	(25·149)	37·920	259·298	340·127
6.	C	11·2564	24·376	47·871	64·476	391·986
7.	N	14·529	29·593	47·426	77·450	97·863
8.	O	13·614	35·108	54·886	77·394	113·873
9.	F	17·418	35·012	{ 62·689 / 62·646	87·139	114·214
10.	Ne	21·559	40·955	(63·450)	97·024	126·260
11.	Na	5·138	47·290	71·628	98·880	138·367
12.	Mg	7·644	15·031	80·117	109·294	141·231
13.	Al	5·984	18·823	28·441	119·957	153·772
14.	Si	8·149	16·339	33·459	45·130	166·725
15.	P	10·484	19·720	30·156	51·354	65·007
16.	S	10·357	23·345	34·799	47·292	(72·474)
17.	Cl	13·014	23·798	39·649	53·450	67·801
18.	A	15·755	27·619	40·705	(75·002)	
19.	K	4·339	31·620	45·793	(60·897)	(82·6)
20.	Ca	6·111	11·868	50·881	(67·181)	84·385
21.	Sc	6·5384	12·797	24·753	73·911	(91·847)
22.	Ti	6·818	13·573	27·467	43·236	99·842
23.	V	6·743	14·651	29·314	(48·464)	65·198
24.	Cr	6·764	16·493	30·950	(48·580)	(73·093)
25.	Mn	7·432	15·636	33·690	(53)	(76·006)
26.	Fe	7·868	16·178	30·643	(56)	(79)
27.	Co	7·875	17·052	33·491	(53)	(82)
28.	Ni	7·633	18·147	35·165	(56)	(79)
29.	Cu	7·724	20·286	37·079	(59)	(83)
30.	Zn	9·391	17·959	39·701	(62)	(86)
31.	Ga	5·997	20·509	30·702	(64·157)	(90)
32.	Ge	7·889	15·930	34·215	45·700	(93·434)
33.	As	9·813	18·628	28·344	50·122	62·612
34.	Se	9·750	21·512	31·979	{ (42·898) / (47·3)	73·103
35.	Br	11·843	21·799	35·887	(50·186)	(59·7)
36.	Kr	13·996	24·565	36·940	(52)	(65·66)
37.	Rb	4·176	{ 27·499 / 27·560	39·664	(53)	(71)
38.	Sr	5·692	11·027	(43·6)	(57·017)	(72)
39.	Y	6·377	12·233	20·514	(61·8)	(76·849)
40.	Zr	6·835	13·126	22·980	34·330	(82 83)
41.	Nb	6·881	14·316	25·038	38·251	50·534
42.	Mo	7·097	16·151	27·133	(46·38)	(61·152)

* After LAKATOS, BOHUS and MEDGYESI (1959) except those given in italics which are either from LANDOLT-BÖRNSTEIN (1950) or FINKELNBURG and HUMBACH (1955). LAKATOS *et al.* list numerous sources (up to mid-1958) for their data. Values in brackets are uncertain or have been obtained through extrapolation and not direct measurement.

Source: Reprinted with permission from L. H. Ahrens, *Physics and chemistry of the earth,* © 1964, Pergamon Books, Ltd.

2.2) and is a much smaller quantity than the ionization energy. The power, or the tendency, of an atom in a molecule to gain or attract electrons is measured by its *electronegativity* χ, most commonly expressed in arbitrary units (Table 2.4). Electronegativity values also increase in the periodic sequence *s, d* or *f, p*. Thus a high ionization potential will be matched by a high electronegativity. Stating this differently, those elements with a low electronegativity will readily

Table 2.3. *(cont.)*

Z	Element	I	II	III	IV	V
43.	Tc	7·276	15·258	29·537	(43)	(59)
44.	Ru	7·364	16·758	28·459	(47)	(63)
45.	Rh	7·461	18·072	31·049	(46)	(67)
46.	Pd	8·334	19·423	32·921	(49)	(66)
47.	Ag	7·574	21·481	34·818 (36)	(52)	(70)
48.	Cd	8·991	16·904	37·766	(55)	(73)
49.	In	5·785	18·865	(28·025)	(54·4 (57·9)	—
50.	Sn	7·342	14·628	30·494	40·724*	72·263
51.	Sb	8·639	16·526 18·593	25·317	44·146	55·691)
52.	Te	9·007	18·593	30·616	37·816	60·270
53.	I	10·454	19·094	(31)	(42)	(52)
54.	Xe	12·127	21·204	32·114	(45)	(57)
55.	Cs	3·893	25·070 23·37	(33·97)	(46)	(62)
56.	Ba	5·210	10·001	(37)	(49)	(62)
57.	La	5·614	11·433	19·166	(52)	(66)
58.	Ce	6·54 6·91	12·31	19·870	36·714	(70)
59.	Pr	(5·76)	—	—	—	—
60.	Nd	(6·31)	—	—	—	—
61.	Pm	—	—	—	—	—
62.	Sm	5·6	11·4	—	(36·5)	—
63.	Eu	5·67	11·24	—	—	—
64.	Gd	6·16	(12)	—	—	—
65.	Tb	(6·74)	—	—	—	—
66.	Dy	(6·82)	—	—	—	—
67.	Ho	—	—	—	—	—
68.	Er	—	—	—	—	—
69.	Tm	—	—	—	—	—
70.	Yb	6·22	12·10	—	—	—
71.	Lu	6·15	14·7	—	—	—
72.	Hf	7·003	(14·874)	(21)	(31)	—
73.	Ta	7·883	(16·2)	(22)	(33)	(45)
74.	W	7·982	(17·7)	(24)	(35)	(48)
75.	Re	7·875	(16·6)	—	—	—
76.	Os	8·732	(17)	—	—	—
77.	Ir	9·1	—	—	—	—
78.	Pt	8·962	18·558	(29)	(41)	(55)
79.	Au	9·223	20·045 20·452	(30)	(44)	(58)
80.	Hg	10·435	18·751	34·210	(46)	(61)
81.	Tl	6·106	20·423	29·822	50·708	(64)
82.	Pb	7·415	15·028	31·929	42·310	68·792
83.	Bi	7·287	16·684	25·556	45·304	56·0
84.	Po	8·426	—	—	—	—
85.	At	—	—	—	—	—
86.	Rn	10·746	—	—	—	—

* Various values of the fourth ionization potential of Sn, ranging from 39·4 to 46·4 V, have been given in the literature. The value given here is close to that 41 V recommended by AHRENS (1956).

transfer an electron to a bonding partner possessed of a high electronegativity; the latter will acquire the electron and its negative charge.

A positively charged atom is called a *cation;* a negatively charged atom, an *anion.* Their union constitutes the *ionic bond.* From Tables 2.3 and 2.4, it follows that those elements having a maximum difference in electronegativity $\Delta\chi$ or in ionization potential ΔI will have the strongest tendency to form ionic bonds: for example, for Na^+, $\chi = 0.9$; for F^-, $\chi = 4.0$ and $\Delta\chi = 3.1$.

Linus Pauling, the great American chemist who developed a still widely used table of electronegativities, or the electronegativity scale, also developed formulae for estimating the percentage of ionic bonding associated with different values of $\Delta\chi$. Thus, for NaF, the value $\Delta\chi = 3.1$ indicates that the compound has 91% ionic bonding. This leads to the concept that bonding of atoms in crystals is often only partly ionic and partly something else. Here, we may briefly categorize the principal chemical bond types:

> the ionic bond, involving *electron transfer*
> the covalent bond, involving *electron sharing*
> the metallic bond, involving *electron mobility*
> the hydrogen bond, involving *electron orientation*
> the van der Waals bond, involving *electron synchronization*

Actually, Pauling developed his concept of electronegativity by determining that the energy holding bonded atoms together in simple chemical compounds is always equal to or greater than the energy of a covalent bond between the constituent atoms. The excess energy is due to the ionic energy of the bond. The line of reasoning is as follows. The simple compound H_2 is fully covalent, and a spherical s orbital of each atom is overlapped so that one electron from each H atom, $1s^1$, is shared (Fig. 2.2). In this s–s overlap, each atom acquires the electronic structure of helium, He, $1s^2$. The bond dissociation energy D equals 104.2 kcal/mole. In Cl_2, also fully covalent, a dumbbell-shaped orbital of each chlorine, Cl, atom, $3s^23p^5$, is overlapped with that of the other so that now the sharing of one electron each from both Cl atoms is diagrammed as in Figure 2.2. In this p–p overlap, each Cl atom acquires the electronic structure of argon, Ar, $3s^23p^6$. The bond dissociation energy D equals 58.0 kcal/mole. Thus, D for the compound HCl should be equal to

$$\tfrac{1}{2}H_2 + \tfrac{1}{2}Cl_2 = 162.2/2 = 81.1 \text{ kcal/mole}$$

Figure 2.2. Orbital overlap in H_2 and Cl_2.

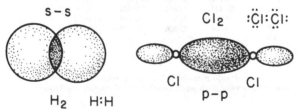

Table 2.4. *Electronegativities of the elements*

Element	Pauling and others as listed by Fyfe (1951)	Little and Jones (1960)	Element	Pauling and others as listed by Fyfe (1951)	Little and Jones (1960)
H	2·1	2·1	Sb	III 1·8 / V 2·1	1·82
He	—	—	Te	2·1	2·01
Li	1·0	0·97	I	2·6	2·21
Be	1·5	1·47	Xe	—	—
B	2·0	2·01	Cs	0·7	0·86
C	2·5	2·50	Ba	0·85	0·97
N	3·0	3·07	La	0·85	1·08
O	3·5	3·50	Ce	1·05	1·08
F	4·0	4·10	Pr	1·1	1·07
Ne	—	—	Nd	—	1·07
Na	0·9	1·01	Pm	—	1·07
Mg	1·2	1·23	Sm	—	1·07
Al	1·5	1·47	Eu	—	1·01
Si	1·8	1·74	Gd	—	1·11
P	2·1	2·06	Tb	—	1·10
S	2·5	2·44	Dy	—	1·10
Cl	3·0	2·83	Ho	—	1·10
A	—	—	Er	—	1·11
K	0·8	0·91	Tm	—	1·11
Ca	1·0	1·04	Yb	—	1·06
Sc	1·3	1·20	Lu	—	1·14
Ti	IV 1·6 / III 1·35	1·32	Hf	(1·3)	1·23
V	IV 1·6 / V 1·8 / II 1·5	1·45	Ta	(1·4)	1·33
Cr	III 1·6 / VI (2·1) / II 1·4	1·56	W	IV (1·6) / VI 2·1	1·40
Mn	III (1·5) / VII (2·3)	1·60	Re	—	1·46
Fe	II 1·65 / III 1·8	1·64	Os	(2·1)	1·52
Co	1·7	1·70	Ir	2·1	1·55
Ni	1·7	1·75	Pt	2·1	1·44
Cu	I 1·8 / II 2·0	1·75	Au	2·3	1·42
Zn	1·5	1·66	Hg	I 1·8 / II 1·9	1·44
Ga	1·6	1·82	Tl	I 1·5 / III 1·9	1·44
Ge	1·7	2·02	Pb	II 1·6 / IV 1·8	1·55
As	2·0	2·20	Bi	1·8	1·67
Se	2·3	2·48	Po	2·0	1·76
Br	2·8	2·74	At	2·4	1·90
Kr	—	—	Rn	—	—
Rb	0·8	0·89	Fr	0·7	0·86
Sr	1·0	0·99	Ra	0·8	0·97
Y	1·2	1·11	Ac	(1·0)	1·00
Zr	1·4	1·22	Th	1·1	1·11
Nb	(1·6)	1·23	Pa	(1·4)	1·14
Mo	IV (1·6) / V (2·1)	1·30	U	1·3	1·22
Tc	—	1·36	Np	—	1·22
Ru	2·05	1·42	Pu	··	1·22
Rh	2·1	1·45	Am	··	(1·2)
Pd	2·0	1·35	Cm	··	(1·2)
Ag	1·8	1·42	Bk	··	(1·2)
Cd	1·5	1·46	Cf	··	(1·2)
In	1·6	1·49	Es	—	(1·2)
Sn	II 1·65 / IV 1·8	1·72	Fm	····	(1·2)
			Md	··	(1·2)
			No ?	···	(1·2)

Source: Reprinted with permission from L. H. Ahrens, *Physics and chemistry of the earth,* © 1964, Pergamon Books, Ltd.

The determined value for $D_{HCl} = 103.2$, and

$$103.2 - 81.1 = 22.1 \text{ kcal/mole}$$

The value $22.1 = \Delta$, the extra ionic contribution or *ionic resonance energy*, and the formula

$$\Delta = D(A - B) - \tfrac{1}{2}D(A - A) + 1/D(B + B)$$

is applicable.

Using bond energies from dielectric moments or measurements, Pauling quantified this excess ionic resonance energy Δ by assigning values of electronegativity χ for each chemical element such that $\Delta\chi$ measures the ionic contribution to bonding between two elements. After assigning a value of $\chi = 2.05$ for H, Pauling derived values of $\chi = 3.0$ for Cl, $\chi = 3.5$ for O, and $\chi = 4.0$ for F, from simple H compounds. He then derived values for χ for other elements by a substitution scheme.

Other investigators have obtained electronegativity values which vary slightly from those of Pauling, and some of these give lower percentages of ionic bonding than those calculated by Pauling, who cautioned that his electronegativity scale was only semiquantitative. Nonetheless, very similar electronegativity values were obtained by Mulliken, who averaged the first ionization potential and the electron affinity of an atom according to the formula $I_1 + E_A/125 = \chi$, when the values for electron affinity and ionization potential are expressed in kcal/mole. The element fluorine, F, has an ionization potential of 17.42 ev; and 1 ev = 23.06 kcal/mole, giving 401.71 kcal/mole for F. The electron affinity value for F is 83.5 kcal/mole. Their sum divided by 125 yields 3.88 for the electronegativity of F, essentially the same value as that obtained by Pauling, who rounded it to 4.0.

It is significant that the following different equations or methods used to estimate the percentage of ionic bonding all give similar, although not identical, results.

For HCl, with $\chi_H = 2.1$ and $\chi_{Cl} = 3.0$, we obtain Pauling's relation

$$\text{Percent ionic bonding} = \Delta/D$$

where $\Delta = 23(\chi_A - \chi_B)^2 = 18.63$ and $D = $ bond dissociation energy $= 103.2 = 18.1\%$, and Hannay and Smith's relation

$$\text{Percent ionic bonding} = 16(\Delta\chi) + 3.5(\Delta\chi)^2 = 17.2\%$$

From dielectric properties and dipole moments, μ:

$$\mu_{obs}/\mu_{calc} \quad \text{for } H^+ - Cl^- = 17.0\%$$
$$(\mu_{obs} = 1.03 \text{ debyes}/\mu_{calc} = 6.07 \text{ debyes})$$

From the foregoing discussion, it should be evident that the chemical bonds joining atoms into crystal structures need not be solely covalent, ionic, or metallic, but are a mixture of these types. In addition to Pauling's criterion of differences in electronegativity, we have other means of evaluating the nature of the chemical bond in crystals.

Measurement of other physical and chemical properties of crystals also provide equally, if not more, valuable criteria for evaluation of chemical bond type. These properties include luster, reflectivity, magnetism, electrical and thermal conductivities, hardness, color, refractivity and index of refraction, coordination number, sums of radii of bonded atoms, solubility of crystals in nonpolar and polar solvents, and the distortion of atomic orbitals in atoms or ions in response to the location of charged atoms or ions in different crystal structure sites.

Thus, although Pauling's electronegativity differences provide the primary and best theoretical method of predicting chemical bond types, measurement of the physical and chemical properties cited provides more direct evidence to confirm those inferences.

Correlation of the physical and chemical properties of minerals with chemical bond type will be developed further in the ensuing chapters.

Summary

When an atom is excited by the addition of external energy, the energy absorbed causes electrons to leap to a higher energy state; when they fall back, they emit the added energy as spectral lines characteristic of the particular element.

Elements with the most loosely bound valence electrons (e.g., s electrons) are easiest to excite.

The multiplicity M of spectral lines is related to the spin quantum number S of the electron ($M = 2S + 1$).

Ionization potential is the energy required to remove an electron from a neutral atom to infinity.

Elements with low ionization potential are most apt to form ionic bonds.

Elements that attract electrons have electron affinity, expressed as electronegativity.

Those elements with low ionization potential also have low electronegativity.

Elements with low electronegativity combine with those having high electronegativity to form ionic bonds most readily.

Although bonds between elements may be ionic, covalent, metallic, or van der Waals, they are most commonly mixtures of more than one bond type.

Percent ionic bonding can be estimated from differences in electronegativity. In addition, bond character can be inferred from physical properties such as luster, hardness, transparency, and electrical conductivity.

Bibliography

Ahrens, L. H. (1953). The use of ionization potentials, Part 2. Anion affinity and geochemistry. *Geochim. et Cosmochim. Acta,* 3: 1–29.

(1964). The significance of the chemical bond for controlling the geochemical distribution of the elements. Part 1, Appendix A. *Physics and chemistry of the earth,* Pergamon, New York, vol. 5, pp. 3–54.

and Taylor, S. R. (1961). *Spectrochemical analysis,* 2d ed. Pergamon Press, New York.

Aller, L. H. (1961). *The abundance of the elements.* Interscience, New York.

Fyfe, W. S. (1964). *Geochemistry of solids.* McGraw-Hill, New York, chaps. 6, 7.

Hannay, N. B., and Smyth, C. P. (1946). The dipole moment of hydrogen fluoride and the ionic character of bonds. *J. Am. Chem. Soc.,* 68: 171–3.

Harrison, G. R. (1939). *M.I.T. wavelength tables.* Wiley, New York.

Herzberg, G. (1944). *Atomic spectra and atomic structure.* Dover, New York.

(1950). *Molecular spectra and molecular structure,* 2d ed. Van Nostrand, New York.

Jaffe, H. W. (1947). Reexamination of sphene (titanite). *Am. Mineral.,* 32: 637.

(1949). Visual arc spectroscopic detection of halogens, rare earths and other elements by use of molecular spectra. *Am. Mineral.,* 34: 667.

Lakatos, B., Bohus, J., and Medgyesi, Gy. (1959). A new way for the calculation of the degree of polarity of chemical bonds. *Acta Chim. Hungar.,* 20: 1; 21: 293.

Little, E. J., and Jones, M. M. (1960). A complete table of electronegativities. *J. Chem. Ed.,* 37: 231.

Merrill, P. W. (1963). *Space chemistry.* Univ. of Michigan Press, Ann Arbor, Mich.

Mulliken, R. S. (1934). A new electronegativity scale; together with data on valence states and on valence ionization potentials and electron affinities. *Jr. Chem. Phys.,* 2: 782.

Pauling, L. (1960). *The nature of the chemical bond.* 3d ed. Cornell Univ. Press, Ithaca, New York, chap. 3.

Pearse, R. W. B., and Gaydon, A. G. (1950). *The identification of molecular spectra,* 2d ed. Wiley, New York.

Peterson, M. J., and Jaffe, H. W. (1953). Visual arc spectroscopic analysis. U.S. Dept. of the Interior, Bur. Mines, Bull. 524, Washington, D.C.

Peterson, M. J., Kauffman, A., and Jaffe, H. W. (1947). The spectroscope in determinative mineralogy. *Am. Mineral.,* 32: 322.

Slavin, M. (1940). Prism versus grating for spectrochemical analysis. Proc. 7th Summer Conf. on Spectroscopy, p. 51.

3 The crystal chemistry of the covalent bond

This chapter considers a variety of crystal structures in which the atoms are bonded by the sharing of one or more electrons. Wave functions ψ of atom pairs are added in such a way that electron density charge clouds overlap to produce a new charge cloud with even greater electron density than the sum of its parts.

Pure or extreme covalent bonds can be formed only between atoms with equal electronegativity, such that $\Delta\chi_{A-B} = 0$, equal numbers of valence electrons, and equal coordination numbers. Obviously, these conditions are fulfilled only in the combination of two atoms of the same species.

$s-s$, $p-p$, and $s-p$ bonds

In the covalent bonding of two hydrogen atoms in H_2, two spherical s orbitals overlap to yield H_2 (Fig. 3.1A).

Using dots for electrons, we may use *Lewis notation* to indicate that the two H atoms are joined by a *single covalent bond* in which one electron from each atom is shared. For H_2, $H \cdot + \cdot H = H \colon H$, and thus each H atom is surrounded by two electrons rather than one, conferring on each H atom the outer or extranuclear electronic structure of helium, $He \colon$, the nearest rare gas. In both covalent and ionic bonding, linked atoms tend to attain the electronic structure of the nearest rare gas and, in so doing, achieve the high stability associated with s^2p^6 completed shells of rare gas atoms. This is known as the *octet rule*.

Similarly, when chlorine, Cl, bonds to Cl in Cl_2, two dumbbell-shaped p orbitals overlap to attain the electronic structure of the nearest rare gas, in this case argon, Ar (see Fig. 3.1B).

In Lewis notation for Cl_2,

$$: \overset{..}{\underset{..}{Cl}} \cdot + \cdot \overset{..}{\underset{..}{Cl}} : \; = \; : \overset{..}{\underset{..}{Cl}} : \overset{..}{\underset{..}{Cl}} :$$

and each Cl atom now has eight rather than seven electrons in its outermost orbital, which is the same as, or *isoelectronic* with, Ar, $_{18}Ar = (_{10}Ne) + 3s^23p^6$, the nearest rare gas.

It has already been shown that the bonding of one atom of H with one of Cl to give HCl has an additional ionic component Δ, and the bond is thus no longer entirely covalent. Nevertheless, the small electronegativity difference $\Delta\chi_{H-Cl} = 2.1 - 3.0 = 0.9$ indicates an ionic contribution of only 19%, and the bonding is dominantly covalent. The electronic overlap of H and Cl is shown in Figure 3.1C. Here, in combination, H has the extranuclear electronic structure of He, $1s^2$, and Cl, the structure of Ar, $3s^23p^6$, obeying the octet rule.

σ, π, and single, double, and triple covalent bonds

All three examples cited earlier – s–s for H_2, p–p for Cl_2, and s–p for HCl – involve single covalent bonds of the type classed as σ bonds; these are symmetrical and rotatable about the bond axis.

In compounds such as the oxygen molecule O_2, two unpaired electrons in each atom are available for bond formation. The orbital configuration of oxy-

Figure 3.1. Covalent bonds: (A) s–s (H_2); (B) p–p (Cl_2); (C) s–p (HCl); (D) p–$\pi(O_2)$.

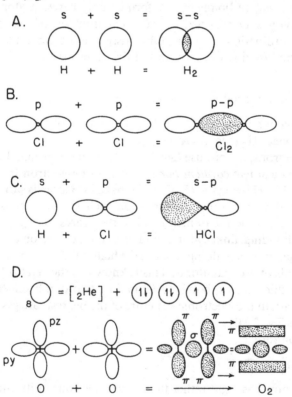

gen is shown in Figure 3.1D. Both the p_y and p_z orbitals may be used to form a *double bond*. In Figure 3.1D, the p_y orbitals along the y bond axis form a σ bond (rotatable), whereas the p_z orbitals combine by "sideways" overlap to form a π bond (nonrotatable or of restricted rotation). The latter is finally further combined into the "stretched rubber-band" configuration shown in Figure 3.1D to become a delocalized orbital that extends throughout the molecule; it is a molecular orbital. Note that the *double bond* molecule consists of a σ bond and a π bond and has drawn the atoms closer together than would a σ bond alone. It should also be apparent that the double bond is stronger than a single bond. Later we will show that the closer the atoms can be brought together, the greater will be the hardness or scratch resistance of the mineral. For this reason alone, minerals containing essential amounts of small atoms with low atomic number, such as boron B, and beryllium, Be, all have great hardness.

Triple covalent bonds also occur when three electrons from each atom are available for bonding. These are usually formed from one σ bond and two π bonds, as in the nitrogen molecule N_2:

$$_7N = [_2He] + \text{①} \; \text{①①①}$$
$$\quad \quad \quad \quad \quad {}_{2s^2} \quad {}_{2p^3}$$

$$:\!\overset{\cdot}{\underset{\cdot}{N}}\!\cdot + \cdot\overset{\cdot}{\underset{\cdot}{N}}\!: \; = \; :N\!:\!:\!:N\!: \quad \text{or} \quad N\!\equiv\!N$$

Here, the three mutually perpendicular orbitals, p_x, p_y, and p_z, will form a triple bond, and each nitrogen atom in N_2 will attain the stable structure of $_{10}Ne$. The shortening of the interatomic distance in going from a single to a double to a triple bond and the consequent increase in bond strength are illustrated in Table 3.1.

The atomic radius for singly bonded carbon, C, 0.77 Å, is obtained by the convention of halving the interatomic distance, and this value is expected to be preserved with close correspondence in many different crystal structures such as diamond, C^{IV}, and silicon carbide, $Si^{IV}C^{IV}$.

Minerals such as diamond, C^{IV}, and graphite, C^{III}, having identical chemical

Table 3.1. *Bond energy, interatomic distance, and atomic radius in single, double, and triple covalent bonds*

Bond	Bond energy (kcal/mole)	Interatomic distance (Å)	Atomic radius (Å)
C—C (single)	83	1.54	0.77
C=C (double)	146	1.33	0.665
C≡C (triple)	200	1.20	0.600
N—N (single)	38	1.48	0.74
N=N (double)	100	1.24	0.62
N≡N (triple)	226	1.10	0.55

composition but different crystal structure are classed as *polymorphs*. The coordination numbers, CN IV for C in diamond and CN III for C in graphite, inform us that the C atoms are more closely packed in diamond than in graphite, and that the former is the high pressure polymorph of carbon.

Promotion and hybridization of covalent bond orbitals

The sp³ hybrid bond orbital

In the ground or unexcited state, $_6$C has the electronic configuration

$$_6\text{C} = (_2\text{He}) + \begin{array}{cccc} \text{⑪} & \text{①} & \text{①} & \text{○} \\ 2s & 2p_x & 2p_y & 2p_z \end{array}$$

indicating that each C atom has four electrons available for bond formation outside of the closed He $1s^2$ shell. In the excited state, these electrons may be used in different ways. Whenever four valence electrons are available, as in C, silicon, Si, tin, Sn, and germanium, Ge, a stronger bond can be formed between atom pairs if all four electrons have equal energy. Thus, in diamond, the four

Figure 3.2. Hybrid bond orbitals, sp^3, and tetrahedra in diamond.

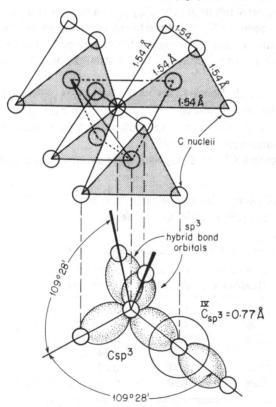

valence electrons are made equal in energy by unpairing one of the $2s$ electrons of C and promoting it to the empty p_z orbital. Thus $_6C = (_2He) + 2s^2 2p^2$ is excited to the configuration $2s^1 2p^3$ and is rewritten as the C_{sp^3} hybrid, as follows:

$$_6C = (_2He) + [①①①①]$$

The four new or hybridized orbitals, now equal in energy, will project four semi-dumbbell-shaped globes of electron density or electron clouds from the C nucleus to the four corners of a regular tetrahedron. This is the classic sp^3 hybrid bond orbital (Fig. 3.2), in which lines drawn from the nucleus through each of the four orbitals will subtend angles of $109°28'$, the regular tetrahedral angle. In diamond, each lobe of the sp^3 hybrid on C will overlap equally with a lobe of an adjoining C atom. This overlap will place paired electrons in each orbital, conferring on each C atom the $2s^2 2p^6$ electron structure of neon. In diamond, we are dealing with a framework extended tetrahedrally in three-dimensional space. Each C atom is at the center of a tetrahedron of C atoms and is at the same time a common corner to four other tetrahedra (Fig. 3.2). In diamond, $8/[C^{IV}]$, the eight carbon atoms of the unit cell or smallest repeat unit form a face centered cube (Fig. 3.3) with the atoms disposed as follows:

8 C on corners $\times \frac{1}{8} = 1$
6 C on face centers $\times \frac{1}{2} = 3$
4 C inside the cube $\times 1 = \underline{4}$
                    8 C atoms in 1 unit cube (see also Fig. 10.1)

If we replace the eight carbon atoms of diamond with four atoms of zinc, Zn, and four atoms of sulfur, S, we derive the structure of sphalerite, 4/ZnS, the principal ore mineral of zinc. Thus, for Figure 3.4, C atoms numbered 2 and 4 in Figure 3.3 represent Zn and those numbered 1 and 3 represent S. The unit

Figure 3.3. Ball model of carbon atoms in diamond.

cell will dispose its eight atoms as follows:

$$8 \text{ Zn at corners} \quad \times \tfrac{1}{8} = 1 \Big\}$$
$$6 \text{ Zn on face centers} \quad \times \tfrac{1}{2} = 3 \Big\} \; 4 \text{ Zn}$$
$$4 \text{ S inside the cell} \quad \times 1 = 4 \quad 4 \text{ S}$$

Although the orbital configuration and valence electron structure of Zn, $3d^{10}4s^2$, and S, $3s^23p^4$, would not suggest an sp^3 hybridization of either atom, both achieve this by pooling the $4s^2$ electrons on Zn with the $3s^23p^4$ electrons on S and dividing them equally among each atom. Whenever eight valence electrons are available for bonding of two atoms and the electronegativity difference is small, an sp^3 hybrid may be used by both atoms and it does not matter if they are supplied equally by each atom, as in diamond or silicon carbide. The electronegativity difference, Δ, for sphalerite from $Zn_{\chi 1.5} - S_{\chi 2.5}$ is 1.0, and the bond is predicted to be dominantly covalent but with 22% ionic character.

The elements in Group IVB of the periodic table all contain four valence electrons outside of closed electron shells, and thus C, Si, Ge, and Sn would be expected to form sp^3 hybrid bonds. This is borne out by the crystal structures of silicon carbide, silicon, germanium, and "gray" tin, in addition to diamond. From Table 3.2 we can see that B-group elements located an equal number of places on either side of Group IVB elements will combine to form sp^3 hybrid

Figure 3.4. Covalent and ionic packing drawings of sphalerite, ZnS.

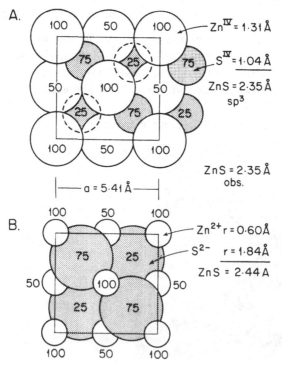

bonds. Further, those compounds having the same sum of atomic numbers have a virtually constant interatomic distance when using this sp^3 hybrid configuration. If the bonds were ionic, the interatomic distances would decrease for compounds having an equal sum of atomic numbers, because of contraction in the radii of cations with increasing charge. These relations are contrasted in Table 3.2, where it is shown that the ionic minerals villiaumite, NaF, and periclase, MgO, both have atomic sums of 20 but show a large difference in interatomic distance. Neither, of course, uses the sp^3 covalent hybrid bond, but rather the close-packed structure of halite, NaCl. The B-group compounds AgI, CdTe, InSb, and SnSn (gray tin) all have atomic number sums of 100, and all have equal interatomic distance sums of 2.80 Å.

It has been shown that the tetrahedral covalent or sp^3 radius of C (0.77 Å) was obtained by halving the distance between the centers of the bonded atoms. Using similar treatment for other covalent compounds, Pauling obtained radii for Si, S, and O from interatomic distances for Si—Si, S—S, and O—O. Then, by a substitution method, he derived a series of single bond tetrahedral covalent radii for use in compounds bonded by sp^3 hybrid orbitals. Radii used by the

Table 3.2. *IV–IV compounds of B-group elements bonded by sp^3 hybrid orbitals*

IB s^1p^0	IIB s^2p^0	IIIB s^2p^1	IVB s^2p^2	VB s^2p^3	VIB s^2p^4	VIIB s^2p^5	Rare gas s^2p^6
		$_5$B	$_6$C	$_7$N	$_8$O	$_9$F	$_{10}$Ne
		$_{13}$Al	$_{14}$Si	$_{15}$P	$_{16}$S	$_{17}$Cl	$_{18}$Ar
$_{29}$Cu	$_{30}$Zn	$_{31}$Ga	$_{32}$Ge	$_{33}$As	$_{34}$Se	$_{35}$Br	$_{36}$Kr
$_{47}$Ag	$_{48}$Cd	$_{49}$In	$_{50}$Sn	$_{51}$Sb	$_{52}$Te	$_{53}$I	$_{54}$Xe

	Valence electrons	Atomic nos.	$A - X$ (Å)	$\Delta\chi$
IV–IV tetrahedral sp^3 hybrid compounds				
CuCl	1 + 7	29 + 17 = 46	2.34	1.2
ZnS	2 + 6	30 + 16 = 46	2.35	1.0
GaP	3 + 5	31 + 15 = 46	2.35	0.5
AgI	1 + 7	47 + 53 = 100	2.80	0.8
CdTe	2 + 6	48 + 52 = 100	2.80	0.6
InSb	3 + 5	49 + 51 = 100	2.80	0.5
SnSn	4 + 4	50 + 50 = 100	2.80	0.0
VI–VI ionic compounds				
NaF	1 + 7	11 + 9 = 20	2.31	2.3
MgO	2 + 6	12 + 8 = 20	2.06	3.1

Note: all have eight electrons per atom pair and a small difference in electronegativity, $\Delta\chi$. Ionic compounds, e.g., $Na^{VI}F^{VI}$, $Mg^{VI}O^{VI}$, have large $\Delta\chi$, use higher CN, and use the close-packed NaCl structure.

tetrahedral compounds of Table 3.2 are given in Table 3.3. Here, it must be emphasized that radii differ with chemical bond type, and even the same atom or ion may have a different radius in different types of crystal structures. For this reason, different sets of radii have been derived for use in comparing covalent, ionic, and metallic minerals and for specific structure types as well.

At a short radial distance from the nucleus of atoms and ions, the electron density is high and is easily associated with or recognized as belonging to a specific atom. However, at the considerably greater radial distances occupied by valence electrons, we are dealing with much lower electron density, a considerable amount of relatively empty space, and the complication of orbital overlap. It is now difficult to say where one atom ends and the other begins. The method of halving the interatomic distance to derive radii is a convenience that is replete with pitfalls and contradictions. Atoms and ions are not beads on a string. It is only by comparison and evaluation of crystal chemical and associated physical and chemical properties of a solid that we gain insight into the nature of the chemical bonding operative in it, and can thus draw conclusions as to the size of atoms. Although the distance between atomic centers or nuclei, the interatomic distance, can be measured with high precision, the partition of this distance into discrete atomic radii is subject to speculation.

Although hydrogen in H_2 and oxygen in O_2 use $s-s$ and $p-p$ single covalent bonds, respectively, it is the sp^3 hybrid that is used by oxygen in the water molecule H_2O. Oxygen hybridizes its six $2s^22p^4$ valence electrons into an unusual sp^3 hybrid tetrahedral orbital in which two lobes contain paired electrons (lone pairs) and the other two lobes contain the single unpaired electrons (Fig. 3.5). Hydrogen atoms, each with one electron, occupy the unpaired lobes of the sp^3 oxygen hybrid, and the H_2O molecule becomes polar: the lone pair side negative, and the H_2 side positive. In water, the octet rule is preserved, because each H has the electronic structure of He and each O that of Ne.

Table 3.3. *Single bond tetrahedral covalent radii*

IB	IIB	IIIB	IVB	VB	VIB	VIIB
		B	C	N	O	F
		0.88	0.77	0.70	0.66	0.64
	Mg	Al	Si	P	S	Cl
	1.40	1.26	1.17	1.10	1.04	0.99
Cu	Zn	Ga	Ge	As	Se	Br
1.35	1.31	1.26	1.22	1.18	1.14	1.11
Ag	Cd	In	Sn	Sb	Te	I
1.52	1.48	1.44	1.40	1.36	1.32	1.28

Source: After Pauling (1960).

The sp² hybrid bond orbital

Returning now to the C atom, we can clearly see that the chemical bonding employed in graphite, C^{III}, is entirely different from that than in diamond. Although diamond, C^{IV}, is colorless, adamantine in luster, a transparent insulator with very high refractive index, and the hardest mineral known, graphite, C^{III}, in contrast, is black, of metallic luster, opaque, a conductor of electricity, and one of the softest minerals known. Thus, whereas diamond is used in expensive jewelry, graphite is used in inexpensive "lead" pencils.

The reasons for this remarkable difference in physical properties come down to the disposition of one electron per atom. Diamond uses an sp^3 hybrid bond orbital, hybridizing all four $2s^2 2p^2$ valence electrons, but graphite hybridizes only three of these electrons and leaves the fourth as a p valence electron outside of the hybrid (Fig. 3.6).

Now, three lobes or orbitals promoted to equal energy lie in a plane and project at angles of 120°, or an equilateral triangle. These sp^2 hybrid C lobes join together in graphite into hexagonal sheets because each lobe is shared (Fig. 3.6). These bonds are σ bonds, and are as strong as those in diamond. The fourth, unhybridized p orbital projects perpendicularly above and below the sp^2 sheets. It draws each pair of C atoms closer together and, in so doing, forms a π bond that becomes delocalized by sideways overlap and extends as a "stretched

Figure 3.5. Chemical bonding in the water (H_2O) molecule.

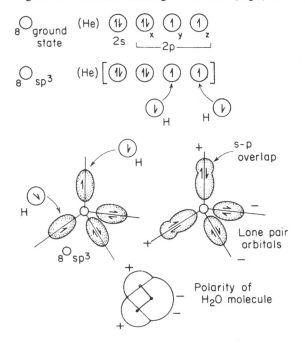

rubber-band" orbital through the entire crystal: it is a molecular orbital (Fig. 3.6). A continuous flow of excited p electrons through this delocalized orbital or band causes absorption of wavelengths of visible light, leading to opacity, metallic luster, and high electrical conductivity.

In addition to σ and π bonds, a third type of weak chemical bond is present – the weak van der Waals bond that holds the graphite sheets together. When we write with a "lead" pencil, we smear sheets of graphite on paper by breaking these weak van der Waals bonds.

The d^2sp^3 hybrid bond orbital

The most common sulfide mineral, pyrite, $Fe[S_2]$ (Figs. 3.7, 3.8), uses a d^2sp^3 hybrid bond orbital on the iron, Fe, atom and an sp^3 hybrid bond orbital on each S atom (Pauling 1960; Burns and Vaughan 1970), and its formula may be written as either $Fe^{VI}[S_2]^{VI}$ or $Fe^{VI}S^{IV}-S^{IV}$. Pyrite uses a variant of the familiar $Na^{VI}Cl^{VI}$ crystal structure: Fe atoms lie on the corners and face centers of the unit cube, and S_2 pairs of dumbbells lie on the twelve edges and in the

Figure 3.6. Chemical bonds, sp^2, π, and van der Waals bonds, in graphite, C.

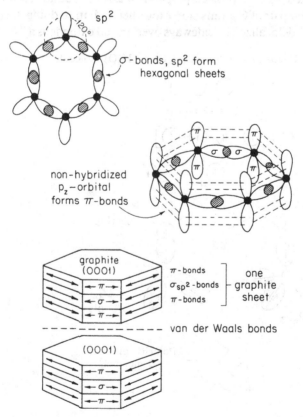

Figure 3.7. Packing models of pyrite with atoms given covalent radii. (A) projected along the c axis [(100) projection] and (B) projected along a threefold axis (111).

Figure 3.8. Packing drawings of the pyrite structure, comparing the ionic with covalent models. Sulfur atoms use sp^3 hybrid bonds in pairs, joining S^I—S^I, and Fe atoms use d^2sp^3 bonds joining Fe^{II} and S_2.

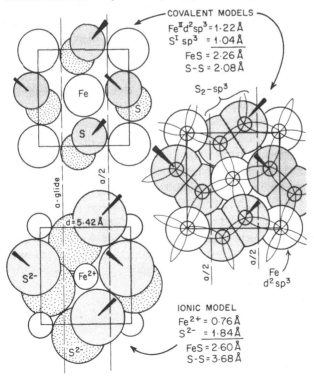

center of the unit cube. It is the different cant or opposed orientation of these S_2 dumbbells that lowers the cubic symmetry of pyrite from the face centered, $Fm3m$ space group of halite to the primitive, $Pa3$ space group. In pyrite, then, S atoms are joined in pairs by sp^3 hybrid bond orbitals by one lobe per atom, and the three remaining lobes overlap and bond with empty d^2sp^3 hybrid lobes of the Fe atom (Figure 3.8). Here, the six $3d$ valence electrons of Fe are paired and placed in three low energy $3d$ orbitals, resulting in a net low electron spin configuration. The two unused high energy $3d$ orbitals, the now empty $4s$ orbital, and three unused $4p$ orbitals are hybridized to equal energy and project six empty lobes toward the corners of a regular octahedron. Now, each Fe atom has loaned its two $4s$ electrons to S. Each Fe atom will be coordinated by VI S atoms, and each S atom will be coordinated tetrahedrally, or in CN IV, by three Fe atoms and one S atom. Thus, each $Fe_{d^2sp^3}$ hybrid has its six empty lobes occupied and overlapped by twelve valence electrons supplied by S atoms. Now, the plot thickens, as the twelve S valence electrons plus the two $4s$ electrons supplied by Fe are believed to join forces to form a fourteen-electron delocalized molecular orbital in which electrons can flow. This explains the metallic luster and electrical conductivity of pyrite. An alternative suggestion by Burns and Vaughan (1970) relates these properties to the formation of a delocalized d-π covalent bond between the paired $3d$ electrons of Fe and the unused d orbitals of S.

Here, we may note that the isostructural compounds, pyrite, FeS_2, and hauerite, MnS_2, although having identical crystal structures, have totally different physical and chemical properties and must therefore have different chemical bonding. $_{25}Mn$ and $_{26}Fe$ differ in the ground state by only one $3d$ valence electron. In both sulfide minerals, the S—S pairs have an identical interatomic distance of 2.16 Å (hence radii of 1.08 Å), but here the similarity ends. Pyrite is totally nonmagnetic or diamagnetic: its electrons fail to orient with an applied magnetic field, whereas hauerite has a paramagnetic response or susceptibility equivalent to five unpaired electrons, its total complement of d electrons. Thus, a magnetic criterion of bond type is apparent and informs us that in pyrite all six Fe $3d$ electrons are paired (low spin state), resulting in a net spin or Bohr magneton value $\mu_B = 0$ (Bohr magneton, $\mu_B = 9.273 \times 10^{-21}$ erg gauss^{-1}), whereas in hauerite all five $3d$ electrons of Mn must be unpaired, yielding $\mu_B = 5.92$. Paramagnetic susceptibility may be measured in units of Bohr magnetons by the formula $\mu_B = \sqrt{n(n+2)}$, where n is the unpaired electron spin per atom (Pauling, 1960). Thus, for the atoms and ions in question, we can derive

$$Fe^{++} \text{ ion,} \quad n = 4, \quad \mu_B = 4.87$$
$$Mn^{++} \text{ ion,} \quad n = 5, \quad \mu_B = 5.92$$
$$Fe^{II} \text{ low spin,} \quad n = 0, \quad \mu_B = 0$$

Pyrite is opaque, metallic, conductive, and diamagnetic, whereas hauerite, although opaque or nearly so, is nonmetallic and strongly paramagnetic. Thus the Mn—S bond must be dominantly ionic, in stark contrast to the d^2sp^3 covalent bond of Fe—S. Thus, the identity of formula and crystal structure does not correctly predict the chemical bond. The electronegativity differences for Fe—S, $\Delta\chi = 0.7$ (= 10% ionic bonding), and for Mn—S, $\Delta\chi = 1.1$ (= 28% ionic bonding), fall short of confirming the dominant ionic character of the Mn—S bond. Evidently, electronegativity, like atomic and ionic radius, is not constant for different crystal structures.

A set of octahedral, d^2sp^3 covalent radii for use in pyrite-type structures is given in Table 3.4. These radii are derived by subtracting half the S—S distance from the metal–S distance (A—X). Because recent determinations of the S—S distance in pyrite (2.14 Å) and in sulfur (2.05 Å) differ from Pauling's assumption that both were 2.08 Å, the values in Table 3.4 may be expected to cover a range and, once again, must be considered to represent approximate values. Although the A—X distances are exact, the radii are not.

The dsp^2 hybrid bond orbital

Not all covalent bonds involving CN IV result from tetrahedral sp^3 hybrids. An interesting group of minerals and synthetic compounds dominated by oxides, chlorides, and sulfides of platinum, Pt, palladium, Pd, nickel, Ni, and copper, Cu, use dsp^2 square (or rectangular) coplanar hybrid bond orbitals in which four lobes of electron density project toward corners of a rectangle, almost a square. In cooperite, PtS (Fig. 3.9), and synthetic PdO, each metal atom pairs eight electrons in four d orbitals and loans its other two valence electrons to either S or O, enabling it to project four empty dsp^2 equal energy hybrid orbitals toward corners of a nearly square rectangle. These are then overlapped by the four S or O atoms at corners of the rectangle. There are now four S atoms around each Pt atom (or four O around each Pd) at corners of a square, and, in turn, four Pt atoms around each S (or four Pd around each O) at corners of a tetrahedron (Fig. 3.9). Note that whenever the bipositive Cu "ion" is coordinated by VI O anions, the four oxygens at corners of a square lie close to the Cu atom, whereas the other two are at a greater distance. The destabiliza-

Table 3.4. *Covalent d^2sp^3 octahedral atomic radii for use in AX_2, pyrite-type, and related structures*

Fe	Cu	Ni
1.17–1.24 Å	1.22–1.33 Å	1.21–1.39 Å
Ru	Rh	Pd
1.28–1.32	1.28–1.32	1.27–1.31
Os	Ir	Pt
1.30–1.34	1.28–1.32	1.27–1.31

Source: After Pauling (1960).

tion of the octahedron in favor of the square suggests that Cu much prefers the *dsp²* hybrid square planar configuration to the ionic octahedron.

The sp linear bond orbital

Earlier it was shown that HCl uses an *sp* linear covalent bond. Only rarely are metallic atoms or ions found coordinated to only two nonmetallic atoms, as in cuprite, Cu_2O, in synthetic Ag_2O, and in the "secondary" uranium, U, minerals containing the uranyl ion $(UO_2)^{2+}$, as in carnotite, a principal uranium ore mineral in Colorado Plateau ores, $K_2(UO_2)_2(VO_4)_2 \cdot 3H_2O$.

In cuprite, Cu_2O, each O atom lies on the corner and center of a body centered cubic lattice, whereas each Cu atom is located on an interpenetrating face centered cubic lattice (see Fig. 18.26). Each Cu atom forms linear, CN II bonds with two O atoms, which are, in turn, each coordinated tetrahedrally by four Cu atoms. The linear O—Cu—O bonds interpenetrate throughout the cubic structure but are not cross-linked (see Fig. 18.26). Electronegativity differences, $\Delta\chi_{Cu-O} = 1.7$, are equivalent to roughly 50% ionic bonding. Cuprite has a deep red color, is barely transparent, and has index of refraction (*n*) of 2.705 (greater than that of diamond), strong internal reflections, and an adamantine luster. Thus, although by electronegativity difference it is 50% ionic, the physical and chemical properties of cuprite are much more those of a covalent rather than an ionic compound.

Although this completes our treatment of the covalent bond, we shall have many further occasions to evaluate the chemical bond in minerals in the ensuing chapters.

Figure 3.9. Rectangular ("square") planar, *dsp²* hybrid bond orbitals on Pt atoms link to four S or O atoms in PtS or PtO (see also Fig. 18.17).

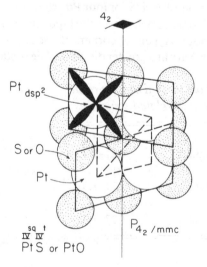

Summary

Covalent bonds are formed by the sharing of electrons. Pure covalent bonds form only between atoms of the same species, by the overlap of s or p orbitals, making single or σ bonds.

Double and triple bonds, stronger than single bonds, are formed of one σ and one or more π bonds. Quadruple bonds may form sp^3 hybrid orbitals, as in diamond, the hardest known compound, and in the water molecule.

In graphite, carbon is linked into sheets by sp^2 (σ) and π bonds; weak van der Waals bonds link the sheets.

In pyrite, Fe atoms use d^2sp^3 hybrid bonds, S atoms use sp^3 hybrid bonds. Hauerite, isostructural with diamagnetic pyrite, is strongly paramagnetic and must therefore be dominantly ionic.

Oxides, chlorides, and sulfides of Pt, Pd, Ni, and Cu use square co-planar dsp^2 hybrid bond orbitals. Linear sp covalent bonds occur in HCl, Ag_2O, Cu_2O, and secondary uranium minerals.

Bibliography

Burns, R. G., and Vaughan, D. J. (1970). Interpretation of reflectivity behavior of ore minerals. *Am. Mineral.,* 55: 1576–86.

Coulson, C. A. (1961). *Valence,* 2d ed. Oxford Univ. Press, London.

Fyfe, W. S. (1964). *Geochemistry of solids.* McGraw-Hill, New York.

Goodenough, J. B. (1972). Energy bands in TX_2 compounds with pyrite, marcasite, and arsenopyrite structures. *J. Solid State Chem.,* 5: 144–52.

Jellinek, F. (1970). Sulfides. *In Inorganic sulfur chemistry,* ed. G. Nickless. Elsevier, Amsterdam, pp. 667–748.

Pauling, L. (1960). *The nature of the chemical bonds,* 3d ed. Cornell Univ. Press, Ithaca, New York, chaps. 4, 5, 6, 7, 11, 14.

Prewitt, C. T., and Rajmani, I. (1974). Electron interactions and chemical bonding in sulfides. *Mineral. Soc. Am. Rev. Mineral.,* 1: 1–4.

Rajmani, V., and Prewitt, C. T. (1973). Crystal chemistry of natural pentlandites. *Can. Mineral.,* 12: 178–87.

Sebera, D. K. (1964). *Electronic structure and chemical bonding.* Blaisdell, Waltham, Mass.

4 The crystal chemistry of the ionic bond

The nature of the ionic bond

Because the covalent bond is directional and a function of spatially oriented electron density clouds, it has been classed as *homopolar*. By contrast, the ionic bond is nonpolar, requires that electric charge be smeared out evenly and symmetrically over the surface of an atom, and is thus classed as *heteropolar*. One atom A of low ionization potential transfers an electron (or more than one) to a bonding partner, atom X, and the cation A^+ and the anion X^- are formed. Atom A shrinks because of the loss of radial distribution and because the nucleus pulls in the remaining electrons more closely to shield its positive charge. Atom X obviously expands enormously, because it adds an outer electron orbiting at a large distance from the nucleus. Electrostatic Coulomb attractive forces will cause A^+ and X^- to approach closely until their valence electrons begin to overlap. At this point, repulsion will set in at a given distance for a given atom pair in a particular crystal structure. This equilibrium distance of closest approach of A^+ and X^- is classed as the *interatomic distance, A — X*, usually cited in ångstrom units, Å ($1 \text{ Å} = 10^{-8} \text{ cm}$).

The ionic model assumes that the A — X distance is equal to the sum of the radii of rigid spheres, A^+ and X^-, in tangent contact. Now, each A^+ cation will attempt to surround itself with the maximum number of X^- anions permitted by geometry, and, in turn, each X^- will surround itself by the maximum number of A^+ ions. The maximum number of X^- ions permitted to surround A^+ is solely a function of the radius of the A^+ cation divided by the radius of the X^- anion, the *radius ratio*. The number of cations A^+ around an X^- anion is a function of the total contribution of electric positive charge to the neutralization of the negative charge of the anion X^-. Fractional charge distribution contributed by each cation to near anion neighbors is equal to cation charge divided by cation coordination number, and is equal to the *electrostatic bond strength* (EBS).

We have just defined the two most important of Pauling's famous five rules for the stability of ionic compounds. We will discuss them in considerable detail later. In ideal ionic bonding, the highest coordination of A^+ around X^- and vice versa results in closest packing of spheres of unlike size and charge; ideal shielding of the electric charge on each partner, equivalent to perfect *ordering;* and a maximum lowering of potential and kinetic energy, contributing to high stability under the particular conditions of formation, such as temperature and pressure.

The ideal ionic bond requires that orbitals of A^+ and X^- not be overlapped. They are thus separated at equilibrium distance by an electron density equal to zero. This density is never achieved, and all so-called ionic minerals have a covalent component, expressed as a significant electron density between bonded atoms or ions.

The magnitude or the percentage of ionic bonding in a compound is predicted by the electronegativity difference, $\Delta\chi_{A-X}$, of bonded atoms, or almost equally well from the difference in the first ionization potential, $\Delta I_{P_{A-X}}$. The closest approach to ideal ionic bonding will occur between atoms of low I_P and high χ. Because low I_P is analogous to low χ, we will continue to use $\Delta\chi_{A-B}$ as our yardstick for prediction of bond type. The lowest electronegativities and I_P values are associated with the alkali elements of periodic group IA or s^1, and the single s^1 valence electron can be removed from Li, Na, K, Rb, and Cs at a very low input of energy of excitation.

Conversely, the atoms having the greatest affinity for gaining electrons in bonding are the halogens of periodic group VIIB, or s^2p^5, only one electron shy

Figure 4.1. Curve relating percentage of ionic bonding to electronegativity difference $\Delta\chi_{A-B}$. Reprinted from Linus Pauling, *The nature of the chemical bond,* 3d ed. Copyright © 1960 by Cornell University. Used by permission of the publisher, Cornell University Press.

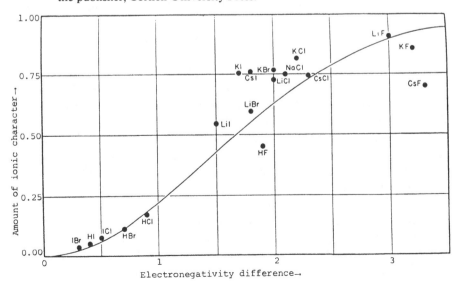

of a stable rare gas configuration. Of these, the electronegativity values fall off rapidly, where $\chi = 4.0, 3.0, 2.8$, and 2.6, for F, Cl, Br, and I, respectively. From the curve of Pauling (1960) relating the percentage of ionic bonding with $\Delta\chi_{A-X}$ (Fig. 4.1), it is apparent that ideal, or nearly ideal, ionic bonds are formed *only* between the alkali ions and fluorine, F, or A^+F^-. Thus, $\Delta\chi_{Na-F} = 3.1 = 91\%$ ionic bonding; $\Delta\chi_{Na-Cl} = 2.1 = 66\%$ ionic bonding; $\Delta\chi_{Na-Br} = 1.9 = 59\%$ ionic bonding; and $\Delta\chi_{Na-I} = 1.7 = 51\%$ ionic bonding. After F, $\chi = 4.0$, comes O with $\chi = 3.5$, and bonds to O with those elements having $\chi \leq 1.7$ will be predominantly ionic with a significant covalent contribution. The Si—O bond, the principal building block of the numerous silicate minerals, has $\Delta\chi = 1.7 = 51\%$ ionic bonding. The Si—O bond may thus be said to resonate between an Si—O ionic bond and an Si—O sp^3 hybrid covalent bond, to which is added an additional $d-\pi$ covalent bond (Brown and Gibbs 1969). The Si—O bond should not be considered to resonate from an ionic bond 50% of the time and a covalent bond 50% of the time, but rather as an ionic–covalent bond 100% of the time. Thus resonance implies that the bond is never wholly ionic nor wholly covalent, but is always a mixture.

The most appropriate generalization of the chemical bond type in minerals is as follows:

fluorides	ionic
chlorides⎫	
oxides ⎬	ionic–covalent
silicates ⎭	
B-group metalloids	covalent
sulfides, arsenides	covalent–metallic
metals, alloys	metallic–covalent

Van der Waals and hydrogen bonds contribute to the ionic–covalent mixture in many minerals.

Because the ionic bond implies and requires sphericity of electron density, it should be obvious that certain valence configurations restrict and prevent formation of spherical ions. All elements of electronic valence configuration s^1 and s^2, alkali and alkaline earth elements of periodic groups IA and IIA will, on ionizing, yield their s^1 and s^2 valence electrons and assume a spherical electron density, as will the halogen, F $2s^22p^5$ of group VIIB on gaining an electron.

Those elements having p, d, or f valence electrons will have spherical electron density only when their threefold, fivefold, and sevenfold orbital energy levels, respectively, are either completely filled or half filled. It is this asymmetry, or departure from sphericity, that favors the formation of covalent bonds and complex anions. Thus, the valence electron structure of the B-group elements – $_5$B, $2s^22p^1$; $_6$C, $2s^22p^2$; $_7$N, $2s^22p^3$; $_{14}$Si, $3s^23p^2$; $_{15}$P, $3s^23p^3$; and $_{16}$S, $3s^23p^4$ – favors formation of covalent complexes with $_8$O, $2s^22p^4$, to produce complex anions of asymmetric electron density. Thus, $[CO_3]^{2-}$, $[NO_3]^{1-}$, $[SiO_4]^{4-}$, $[PO_4]^{3-}$, and $[SO_4]^{2-}$, to name a few, form complex anions that retain their identity and asymmetry in bonding to other elements. Thus, $[CO_3]^{2-}$ is a planar anion rather than a C^{4+} cation bonded to three O^{2-} anions, and $[PO_4]$ is a tetrahedral complex anion rather than a P^{4+} cation bonded to four O^{2-} anions.

We have already seen, in the previous chapter, that an S^{2-} anion does not exist in pyrite, $Fe(S—S)$, but occurs as an $[S_2]^{2-}$ coupled anion complex using an sp^3 hybrid. The existence of the S^{2-} anion should always be regarded with suspicion, inasmuch as S has $\chi = 2.5$ and will, in any mineral, always form bonds that are more covalent than ionic.

Pauling's five rules for coordination (ionic) compounds

The principles that underlie the formation and stability of ionic or coordination compounds were codified by Pauling (1960) into five rules. We will now discuss them in detail and also consider them in subsequent chapters.

Rule 1. *Cation coordination and the radius ratio.* A polyhedron of spherical anions is grouped around each spherical cation such that the number of anions that may surround the cation is a function of the radius ratio $R = r_{cat}/r_{an}$. We assume that each cation is in tangent contact with each anion and lies in the center of the coordination polyhedron of anions that surround it.

Rule 2. *Electrostatic bond strength and the number of polyhedra with a common corner.* The number of cations that may surround a given anion is limited by the requirement that negative electric charge on anions be satisfied locally or over short range by *near cation neighbors*. A given anion will be the common corner to several coordination polyhedra, the exact number limited by the concept of EBS, equal to cation charge/cation CN. We assume that the charge of the cation is distributed evenly among all neighboring anions, and the sum of the fractional EBS contributions reaching an anion from all neighboring cations must equal the valency of the anion with sign reversed. Typical EBS values for some common cations are Li^+/CN VI $= +\frac{1}{6}$, Be^{2+}/CN IV $= +\frac{1}{2}$, Al^{3+}/CN IV $= +\frac{3}{4}$, Si^{4+}/CN IV $= +1$, and Nb^{5+}/CN VI $= +\frac{5}{6}$. Thus, EBS $= +2$ for O^{2-} and $+1$ for F^- or Cl^-. For low or α-quartz, $Si^{IV}O_2^{II}$, each Si^{4+} ion will occupy a tetrahedron of O^{2-} anions, and each Si will contribute $+4/IV = +1$ EBS units, limiting the CN of the O^{2-} anion to II. For halite, $Na^{VI}Cl^{VI}$, each Na^+ inside an octahedron, CN VI, contributes $+\frac{1}{6}$ EBS units to each Cl^-, requiring that there be six Na^+ around each Cl^-, and this leads to a so-called VI–VI structure. Most mineralogy textbooks fail to stress – indeed, they even ignore – the coordination or CN of the *anion:* this is as important as the CN of the cation for interpreting the structure of crystals. A detailed discussion and elaboration of this second rule is given in Chapter 5.

Rule 3. *The rule of polyhedral sharing.* Sharing of corners, edges, and faces of coordination polyhedra brings centers of positive charge closer together than they would be if edges were unshared, and this tends to decrease the stability of coordination compounds. Such sharing, however, is widespread in minerals; otherwise most of these would not be able to polymerize or link together to build extended crystal lattices. Pauling observed that whenever two cations approach one another across a common polyhedral edge, the two anions forming the common edge will draw closer together, reducing their interatomic distance, to better shield their negative charge clouds. This is known as the *shortening of shared edges* and provides proof of the presence of charged atoms in an ionic bond.

Conversely, those minerals in which shared polyhedral edges do not shorten provide evidence of a predominantly covalent bond. Many oxide and silicate minerals in which the O^{2-} anions are in close packing reveal O—O distances of 2.80 Å, twice the radius of O^{2-}, but any two O^{2-} anions forming a shared polyhedral edge show O—O interatomic distances of 2.50 Å, a considerable reduction. Shortening of shared polyhedral edges obviously distorts the geometry of CN polyhedra, and distorted polyhedra are actually the rule, rather than the exception, in minerals (see Chapter 8).

Rule 4. *Independence of polyhedra with small cations of high charge.* Cations with small radius and large charge tend not to share polyhedral elements with one another. This is an extension of Rule 2, where we have seen that in α-quartz, sharing of polyhedra is limited to corners because of the large EBS value of Si^{4+}/IV, which is equal to $+1$, or one-half the negative valence of the O^{2-} anion. Where the nominal electric charge exceeds 4^+, polyhedra containing these cations will remain independent of one another. Thus, EBS values for nominal C^{4+}, P^{5+}, and S^{6+} ions in CN III, IV, and IV, respectively, are $+1.33$, $+1.25$, and $+1.5$. Two such cations around an anion would overcharge the anion, and this restricts polyhedral sharing. Because the isolated tetrahedral $[PO_4]^{3-}$ groups must be joined together by other cations to form a phosphate mineral, it follows that they must be linked by cations with lower EBS. Thus, in berlinite, $[Al^{IV}P^{IV}O_4^{II}]$, isostructural with α-quartz, $[Si^{IV}O_2^{II}]$, each O^{2-} anion may be considered to receive an EBS contribution of P^{5+}/IV $= +1.25$ and Al^{3+}/IV $= +.75 = +2$.

This rule may be somewhat diminished by the fact that C^{4+}, P^{5+}, and S^{6+} cations do not really exist as such, but rather form complex anion groups or complex anions such as $[CO_3]^{2-}$, $[PO_4]^{3-}$, and $[SO_4]^{2-}$. These combine in minerals as though they were single or simple anions of nonspherical shape. Thus, for calcite, we may write the formula $Ca^{VI}C^{III}O_3^{II}$ or $Ca^{VI}[CO_3]^{VI}$. We may illustrate these alternatives in EBS diagrams as follows:

$$EBS = 0.33 + 0.33 + 1.33 = +2 \quad \text{or} \quad 0.33 \times 6 = +2$$

As can be seen here, there are two ways to do the arithmetic, but the second choice is the more realistic. In this connection, Pauling's rules may often be satisfied for covalent as well as for ionic compounds, as can be seen in the above. Here, although calcite is considered to be an ionic mineral, the $[CO_3]^{2-}$ group is itself a covalently bonded complex anion. The fact that no two $[CO_3]$, $[PO_4]$, or $[SO_4]$ polyhedral anion groups ever share corner oxygens with one another in any carbonate, phosphate, or sulfate mineral underscores the validity of this rule.

Rule 5. *The rule of parsimony.* This, the least important of Pauling's rules, states that a large number of different kinds of coordination polyhedra in a given mineral tends to decrease its stability, because this would build unnecessarily complex molecules. In most minerals, only two or three different types of polyhedra are found. Many exceptions to this rule may be cited, most notably the following:

Tourmaline

$$Na^{IX}Mg_3^{VI}Al_6^{VI}[Si_6^{IV}O_{18}][B^{III}O_3]_3[(OH), F]_4$$

and

Hornblende

$$Na_{0.5}^{X}Ca_2^{VIII}(Fe^{2+}, Al)_2^{VI}Fe_2^{2+VI}Mg^{VI}[Si_{6.5}^{IV}Al_{1.5}^{IV}O_{22}][(OH),F]_2$$

Tourmaline contains five different polyhedra, and hornblende seven, yet each exists over a wide range of temperature and pressure in the presence of H_2O and F-bearing fluids.

Radius ratios and coordination numbers in ionic compounds

Atoms widely separated in space or disordered can lower their potential and kinetic energies by bonding together to form ordered crystal structures. We began this book with an analogy of shaking a crate of oranges to bring them into a closer-packed and hence better-ordered state commensurate with lower energy and higher stability. A better analogy would be provided by a crate of tennis and Ping-Pong balls. If we vigorously shake the contents of this box, the tennis balls will pack so that each is surrounded or coordinated by twelve other tennis balls, whereas the much smaller Ping-Pong balls will occupy the interstices or voids formed between the packed tennis balls.

First, let us look at the packing of tennis balls alone. If we shake the box of tennis balls alone, each will wind up in CN XII. One of two arrangements of CN XII is possible: cubic close packing of spheres of equal radius, or hexagonal close packing of spheres of equal radius. *Both are equally close packed* (Fig. 4.2).

First we will pack six tennis balls in a plane around a central tennis ball. Next we will place another equilateral triangle of three tennis balls directly over the central ball of the first layer in close packing. Now we will place another equilateral triangle of three tennis balls directly under the central ball. There are now twelve spheres of equal radius around the central sphere, and we have a close-packed arrangement. However, we may orient the two triangular groups above and below our central sphere in parallel arrangement so that they form a trigonal prism and a hexagonal close-packed array (Fig. 4.2), or, on the other hand, we may orient the two triangular groups above and below the central sphere in opposition so that they form an octahedron, in which case we have a cubic close-packed arrangement (Fig. 4.3). In either case, the voids or interstices between the close-packed tennis balls, each in CN XII, will be of two types. One will be a void enclosed by six tennis balls, a CN VI site, and the other a void enclosed by four tennis balls, a CN IV site. These types will be oriented in alternating rows through the close-packed arrays (Fig. 4.3). It is important to see that the central tennis ball (Fig. 4.3B layers) is now surrounded by fourteen voids: six are octahedral, with three sites above and three below the central sphere; eight are tetrahedral, with four above and four below the same central sphere. In Figure 4.3, we have filled all fourteen sites with CN VI cations of large radius, and CN IV cations of smaller radius. This means that the central sphere, whether a tennis ball or a Cl^- anion, is a corner that is common to all fourteen polyhedra. If the central sphere is designated as a Cl^- anion, it should be

Figure 4.2. Cubic and hexagonal close-packing models and their geometries. Layers stack in sequence, C over B, over A.

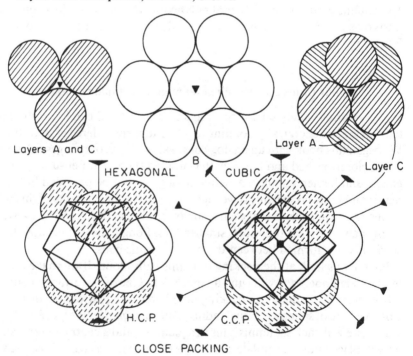

CLOSE PACKING

obvious that occupancy of all fourteen cation sites would convey a considerable excess of positive charge to the central Cl^- anion. Thus, in a close-packed array of anions, only a maximum of six of the fourteen sites will be occupied by cations in any given mineral. The limit on the number of sites that may be occupied by cations is strictly established by Pauling's second rule of electrostatic bond strength, which will be elaborated in considerable detail in the next chapter.

If the array of tennis balls is hexagonally closely packed, a CN IV site will lie directly over another such site, and a CN VI site will also lie directly over another identical site. If the array of tennis balls is cubic close packed, a CN IV site will lie directly over a CN VI site, and vice versa. To state this in another way, in HCP (hexagonal close packing) a tetrahedral site lies above a tetrahedral site, and an octahedral site lies over another of the same kind. In CCP (cubic close packing) the sites alternate positions so that a tetrahedral site lies over an octahedral site, and vice versa. These two arrangements favor the location of

Figure 4.3. Hexagonal close packing (HCP) resulting from placement of layer B over layer A (upper) stacks face sharing octahedra over one another along the threefold symmetry axis, whereas cubic close packing (CCP) with layer B over layer A (lower) stacks octahedra and tetrahedra in alternation along the threefold axis. Constraints on distribution of electric charge favor HCP in corundum, Al_2O_3 (four of six octahedral sites occupied), but favor CCP in periclase, MgO (six of six octahedral sites occupied). In both minerals tetrahedral sites are vacant.

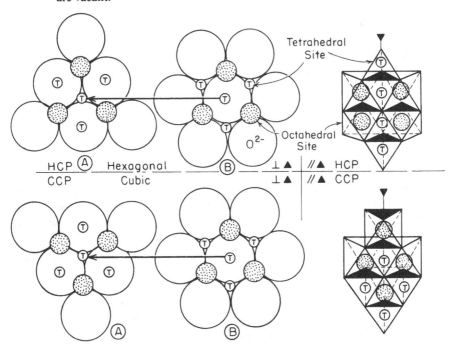

ions differently in different crystal structures, because one or the other arrangement will allow for a more geometrically equalized spacing of cations around an anion. Let us, for example, take the case of halite, NaCl, where our tennis balls will represent large Cl^- anions, and our Ping-Pong balls will represent the smaller Na^+ cations. In halite, each Cl^- is in CCP with XII Cl^- anions around each other. The Na^+ cations are too large to occupy the tetrahedral voids in the Cl^- array, so these will remain empty. The Na^+ cations all must thus occupy the CN VI octahedral sites or voids. Our central Cl^- anion will now be surrounded or coordinated as well by six Na^+ cations at corners of an octahedron: three above are opposed by three below. There are now VI Cl^- around each Na^+ and VI Na^+ around each Cl^-. If we change our Cl^- packing from CCP to HCP, we can still place VI Na^+ around each Cl^- anion, *but* our six points of positive charge, the Na^+ ions, will now be positioned directly above and below each other and the central Cl^- anion, outlining a trigonal prism rather than an octahedron of positive charge. The six points of positive charge will thus be closer together in HCP than in CCP. The CCP array is thus favored for NaCl because it permits more equidistant geometric placement of the six Na^+ cations. For this reason halite crystallizes in the cubic rather than in the hexagonal system. Halite is a VI : VI structure, because each ion is in CN VI.

By contrast, the mineral bromellite, BeO, crystallizes in the hexagonal crystal system. Each O^{2-} anion is in HCP, and now, because Be^{2+} is a very small ion, it will occupy only the tetrahedral, CN IV, sites or voids, leaving the CN VI sites empty. Electrical neutrality requires that only four bipositive Be ions be placed around each O^{2-} anion. The HCP of O^{2-} requires the placement of the four Be^{2+} ions at corners of a tetrahedron, with three below and one above the central O^{2-} anion. In this arrangement, all four BeO_4 tetrahedra point in the same direction, away from the viewer, and this gives rise to a polar structure. If the O^{2-} anions used CCP, the four Be^{2+} cations could not be placed geometrically equidistant, and one of the four would have to be placed too close to another. The hexagonal close-packed polar tetrahedral array is thus favored for bromellite, in contradistinction to halite. Bromellite is therefore a IV : IV structure, because cation and anion are each surrounded by four ions of the opposite species and charge.

Let us now return to our tennis and Ping-Pong balls to take a closer look at the radius ratios that control the site occupancies of the latter. We will start with a close-packed array of tennis balls. It is now immaterial whether we use CCP or HCP. The Ping-Pong ball has a radius of 1.9 cm, the tennis ball 3.2 cm. Thus the radius ratio is $R = r_{PB}/r_{TB} = 1.9/3.2 = 0.594$.

By simple geometry, we can predict that for $R = 0.594$, the Ping-Pong balls will occupy the six-fold CN sites and be in tangential contact with six tennis balls placed at corners of a regular octahedron (Fig. 4.4). Geometric relations show that it is solely the radius ratio R that governs the fit of small spheres in spherical voids or interstices of more or less close-packed large spheres. The maximum number of large spheres that may be in contact with a small sphere is controlled by the minimum radius ratio for a given geometry. The central cation must be in contact with all surrounding anions and must not "rattle" in the void, or else it must adopt a smaller CN polyhedron. The minimum radius ratios for the principal coordination polyhedra of ionic structures, and their geometric derivation, are shown in Table 4.1 and illustrated in Figure 4.4.

Note that it is immaterial whether the large spheres are tennis balls, $r = 3.2$ cm, O^{2-} anions, $r = 1.40$ Å, or Cl^{1-} anions, $r = 1.81$ Å; it is solely the ratio of the small sphere to the large sphere that counts, $R = r_A/r_X$, where A is a small cation and X is a large anion. Note also that for all of these geometrically imposed limitations of occupancy of space we assume that we are dealing with rigid spheres of constant radius in the ionic model. Later it will become apparent that this treatment loses credibility because anions, like tennis balls, are indeed deformable and compressible. For now, we shall ignore or table this complexity, and deal only with our concept of spherical ions of constant radius in a given coordination polyhedron. We will now leave behind the useful tennis

Figure 4.4. Coordination polyhedra, their geometry, and limiting or minimum radius ratios, R. Large spheres are O^{2-} anions, $r = 1.40$ Å.

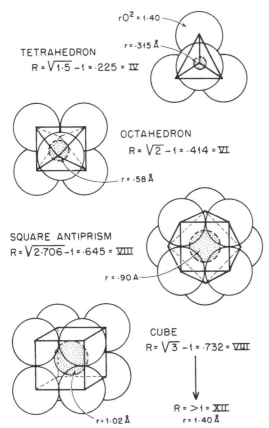

and Ping-Pong ball analogy in order to turn our attention to anions and cations and the derivation of their ionic radii.

Ionic radii

X-rays were discovered by Röntgen in 1895; the diffraction of X-rays by crystals was discovered by Friedrich, Knipping, and Laue in 1912; and the nature of the characteristic X-ray spectra was found by Mosely in 1913, shortly before his untimely death in World War I. Very soon thereafter, W. H. Bragg, W. L. Bragg, and others investigated and published the results of a number of crystal structure determinations based on X-ray analyses. Crystal chemistry was indeed born, and for the first time scientists had some concrete information on the arrangement and location of atoms and ions in crystals. It was not long before physicists, chemists, and geologists began to apportion the X-ray–determined interatomic distances into the radii of spherical ions in tangential contact.

Note that an ion in space has no fixed radius, because the electron cloud of the outermost electron falls off steadily with increasing distance from the nucleus. In crystals, however, in the ideal ionic case (e.g., Na^+F^-), two ions of opposite charge are attracted by Coulomb forces until the close approach of their electron clouds results in rapid repulsion. The balancing of attractive and repulsive forces at some equilibrium distance for a given pair of ions in a given crystal is equal to the $A—X$ or interatomic distance, which, ideally, is the sum of the radii of each ion. The ions are believed to retain these radii in chemical combination with other elements in an additive sense. Recall, however, that electronegativity differences, $\Delta\chi$, for NaF and NaCl are 3.1 and 2.1, equivalent to bonds that are about 10% and 20% covalent, respectively. The orbitals are indeed overlapped, and the apportionment of interatomic distances, 2.31 Å ($Na—F$) and 2.76 Å ($Na—Cl$), into discrete radii for Na^+ = 0.95 Å, for F^- = 1.36 Å, and for Cl^- = 1.81 Å, separated by a zero electron density, is a convenient distortion of the truth.

Table 4.1. *Minimum radius ratio and limiting geometry of regular coordination polyhedra*

Minimum radius ratio R	CN	Polyhedron	Geometry of limit
0.225	IV	Tetrahedron	$R = \sqrt{1.5} - 1$
0.414	VI	Octahedron	$R = \sqrt{2} - 1$
0.645	VIII	Square antiprism	$R = \sqrt{2.71} - 1$
0.732	VIII	Cube	$R = \sqrt{3} - 1$
1.0	XII	Cube–octahedron or hexagonal prism	$R = \sqrt{1}$

One of the first important derivations and compilations of the radii of ions was given by Wasastjerna in 1923. He apportioned interatomic distances on the basis of the molar refractivities of ionic salts in solution, on the assumption that molar refractivities were proportional to ionic volumes. He gave a value of 1.32 Å for the O^{2-} anion and 1.33 Å for the F^- anion. Goldschmidt used these anion values as a basis for the formulation of a fairly complete set of empirical ionic radii, which he published in 1926 (Table 4.2).

Pauling suggested that values of 1.40 Å for O^{2-} and 1.36 Å for F^- were more realistic, and in 1927 published a revised set of ionic radii based on an inverse variation of effective nuclear charge for isoelectronic elements of the alkali halides. He noted that the size of an ion is determined by the radial distribution of the outermost electron, and is inversely proportional to the effective nuclear charge acting on this electron. Effective nuclear charge Z' is equal to the proton number Z minus the screening constant S: $Z' = Z - S$. The screening constant S expresses the degree to which the inner electrons screen the nucleus of the ionized atom. Isoelectronic ions such as $_9F^-$ and $_{11}Na^+$ have the electronic structure of $_{10}Ne$, which is $1s^2 2s^2 2p^6$ and have an electronic screening constant $S = 4.52$. The interatomic distance in $NaF = 2.31$ Å. By division of this distance in the inverse ratio of the effective nuclear charge, Pauling derived ionic radii for Na^+ and F^- as follows:

$$\frac{r_{Na^+}}{r_F} = \frac{9 - 4.52}{11 - 4.52} = \frac{4.48}{6.48} = \frac{41\%(2.31)}{59\%(2.31)} = \frac{0.95 \text{ Å}}{1.36 \text{ Å}}$$

The radii are assumed to be additive, and thus subtraction of the radius of $Na^+ = 0.95$ from the interatomic distance for $NaCl = 2.76$ yields the value of the radius of 1.81 for Cl^-. Continuation of this substitution process leads to derivation of values for other ions.

In 1952, Ahrens revised Pauling's values by using different values for the alkali ions and by plotting regularities in their radii against ionization potentials. On this basis Ahrens provided the most complete table of ionic radii available, one that has been widely used by geochemists and mineralogists (Table 4.2).

In 1964, Fumi and Tosi derived a new set of ionic or "crystal" radii (Table 4.2) for the alkali halides stated to have a very high degree of accuracy. These radii were based on Born repulsive parameters and electron density plots. The Fumi and Tosi radii are startling, in that they indicate considerably larger values for cation radii and correspondingly smaller values for anion radii than do any of the previous sets. If cations are larger and anions are smaller than previously indicated, then the concept of the radius ratio would appear to be invalid: individual bonds would then be more important than packing considerations in determining site occupancies. Perhaps more work with electron density plots will help resolve this major inconsistency.

In 1969, Shannon and Prewitt derived another set of "effective" ionic radii stated to represent most accurately the values found in crystals (Table 4.2).

Table 4.2. Ionic radii by different authors for different coordinations

Ion	E.C.	C.N.	G	A	P	S-P	W-M	F-T
${}_3$Li$^+$	$1s^2$	IV	-	-	-	.59	.68	.73
		VI	.78	.68	.60	.74	.82	.88
${}_4$Be^{2+}	$1s^2$	IV	.34	.35	.31	.27	.35	.31
${}_5$B^{3+}	$1s^2$	III	-	.23	.20	.02	.10	.16
		IV	-	-	-	.12	.20	.26
${}_8$O^{2-}	$2p^6$	II	-	-	-	1.35	1.26	1.21
		III	-	-	-	1.36	1.28	1.22
		IV	-	-	-	1.38	1.30	1.24
		VI	1.32	1.40	1.40	1.40	1.32	1.26
${}_9$F$^-$	$2p^6$	II	-	-	-	1.285	1.21	1.145
		III	-	-	-	1.30	1.22	1.16
		IV	-	-	-	1.31	1.23	1.17
		VI	1.33	1.36	1.36	1.33	1.25	1.19
${}_{11}$Na$^+$	$2p^6$	VI	.98	.97	.95	1.02	1.10	1.16
		VII	-	-	-	1.13	-	1.27
		VIII	-	-	-	1.16	1.24	1.30
		IX	-	-	-	1.32	-	1.46
${}_{12}$Mg^{2+}	$2p^6$	IV	-	-	-	.58	.66	.63
		VI	.78	.66	.65	.72	.80	.86
		VIII	-	-	-	.89	.97	1.03
${}_{13}$Al^{3+}	$2p^6$	IV	-	-	-	.39	.47	.53
		V	-	-	-	.48	-	.62
		VI	.57	.51	.50	.53	.61	.67

Ion	E.C.	C.N.	G	A	P	S-P	W-M	F-T
${}_{24}$Cr^{3+}	$3d^3$	VI	.64	.63	.69	.615	.70	.755
${}_{25}$Mn^{2+}	$3d^5$	VI	.91	.80	.80	.82	.91	.96
		VIII	-	-	-	.93	1.01	1.07
Mn^{3+}	$3d^4$	VI	.70	.66	.66	.65	-	.79
Mn^{4+}	$3d^3$	VI	.52	.60	.54	.54	.62	.68
${}_{26}$Fe^{2+}	$3d^6$	IV	-	-	-	.63	.71	.77
		VI	.83	.74	.76	.77	.86	.91
		VIII	-	-	-	(.91)	-	-
${}_{26}$Fe^{3+}	$3d^5$	IV	-	-	-	.49	.57	.63
		VI	.67	.64	.64	.645	.73	.785
${}_{27}$Co^{2+}	$3d^7$	VI	.82	.72	.74	.735	.83	.875
Co^{3+}	$3d^6$	VI	-	.63	.63	.61	.69	.65
${}_{28}$Ni^{2+}	$3d^8$	VI	.78	.69	.72	.70	.77	.84
Ni^{3+}	$3d^7$	VI	-	-	.62	.60	-	.74
${}_{29}$Cu$^+$	$3d^{10}$	IIs-p	-	-	1.18	.46	-	.60
		VI	-	-	.96	-	-	-
Cu^{2+}	$3d^9$	IVsq	-	-	-	.57	-	-
		V	-	-	-	.65	-	.76
		VI	-	.72	-	.73	-	.79
${}_{30}$Zn^{2+}	$3d^{10}$	IV	-	-	-	.60	.68	.74
		VI	.83	.74	.74	.745	.83	.885
${}_{31}$Ga^{3+}	$3d^{10}$	IV	-	-	-	.47	-	.61
		V	-	-	-	.55	-	.69

Ion	Config	CN						
$_{14}Si^{4+}$	$2p^6$	IV	–	–	–	.26	.34	.40
		VI	.39	.42	.41	.40	.48	.54
$_{15}P^{5+}$	$2p^6$	IV	.34	.35	.34	.17	.25	.31
$_{16}S^{2-}$	$3p^6$		1.74	1.84	1.84	1.72	1.72	–
$_{17}Cl^{-}$	$3p^6$		1.81	1.81	1.81	1.72	1.72	–
$_{19}K^{+}$	$3p^6$	VI	1.33	1.33	1.33	1.38	1.46	1.52
		VIII	–	–	–	1.51	1.59	1.65
		IX	–	–	–	1.55	–	1.69
		X	–	–	–	1.59	1.67	1.73
		XII	–	–	–	1.60	1.68	1.74
$_{20}Ca^{2+}$	$3p^6$	VI	1.06	.99	.99	1.00	1.08	1.14
		VII	–	–	–	1.07	–	1.21
		VIII	–	–	–	1.12	1.20	1.26
		IX	–	–	–	1.18	–	1.32
		X	–	–	–	1.28	1.36	1.42
		XII	–	–	–	1.35	1.43	1.49
$_{21}Sc^{3+}$	$3d^1$	VI	.83	.81	.81	.73	.83	.87
		VIII	–	–	–	.87	.95	1.01
$_{22}Ti^{3+}$	$3d^1$	VI	.69	.76	–	.67	–	.81
Ti^{4+}	$3p^6$	VI	.64	.64	.68	.605	.69	.745
$_{23}V^{3+}$	$3d^2$	VI	.65	.74	.74	.64	.72	.78
V^{4+}	$3d^1$	VI	.61	.63	.60	.59	.62	.73
V^{5+}	$3p^6$	IV	–	–	–	.355	.44	.495
		V	–	–	–	.46	–	.60
		VI	.59	.59	.59	.54	.62	.68
$_{32}Ge^{4+}$	$3d^{10}$	IV	–	–	–	.40	–	.54
		VI	.62	.62	.62	.53	.68	.76
$_{34}Se^{2-}$	$4p^6$		1.91	1.96	1.95	1.98	1.88	–
$_{35}Br^{-}$	$4p^6$		–	1.95	1.95	1.96	–	–
$_{37}Rb^{+}$	$4p^6$	VI	1.49	1.47	1.48	1.49	1.57	1.63
		VIII	–	–	–	1.60	1.68	1.74
		XII	–	–	–	1.73	1.74	1.87
$_{38}Sr^{2+}$	$4p^6$	VI	1.27	1.12	1.13	1.16	1.21	1.30
		VIII	–	–	–	1.25	1.33	1.39
		X	–	–	–	1.32	1.40	1.46
		XII	–	–	–	1.44	1.48	1.58
$_{39}Y^{3+}$	$4p^6$	VI	1.06	.92	.93	.89	.98	1.03
		VIII	–	–	–	1.015	1.10	1.155
		IX	–	–	–	1.10	–	1.24
$_{40}Zr^{4+}$	$4p^6$	VI	.87	.79	.80	.72	.80	.86
		VIII	–	–	–	.84	.92	.98
$_{41}Nb^{5+}$	$4p^6$	IV	–	–	–	.32	–	.46
		VI	.69	.69	.70	.64	.70	.78
$_{42}Mo^{4+}$	$4d^2$	VI	.68	–	–	.65	.73	.79
Mo^{6+}	$4p^6$	IV	–	–	–	.42	.50	.56
		VI	–	–	.62	.60	.68	.74
$_{44}Ru^{4+}$	$4d^4$	VI	.65	.62	.62	.62	.68	.76
$_{45}Rh^{3+}$	$4d^6$	VI	.68	.67	.62	.62	.665	.805
Rh^{4+}	$4d^5$	VI	–	–	–	.615	–	.775

Table 4.2. (cont.)

Ion	E.C.	C.N.	G	A	P	S-P	W-M	F-T
$_{46}Pd^{2+}$	$4d^8$	VI	-	.80	.86	.86	-	1.00
Pd^{4+}	$4d^6$	VI	-	.65	-	.62	-	.76
$_{47}Ag^+$	$4d^{10}$	II	-	-	1.39	.67	-	.81
		VI	1.13	1.26	1.26	1.15	1.23	1.29
		VIII	-	-	-	1.07	1.15	1.21
$_{49}In^{3+}$	$4d^{10}$	VI	.92	.81	.81	.79	-	.93
		VIII	-	-	-	.92	-	1.06
$_{50}Sn^{4+}$	$4d^{10}$	VI	.74	.71	.71	.69	.77	.83
$_{51}Sb^{5+}$	$4d^{10}$	VI	.60	.62	.62	.61	.69	.75
$_{52}Te^{2-}$	$5p^6$	VI	2.11	-	2.21	-	-	-
$_{53}I^-$	$5p^6$	VI	2.20	2.16	2.16	-	-	-
$_{55}Cs^+$	$5p^6$	VI	1.65	1.67	1.69	1.70	1.78	1.84
		IX	-	-	-	1.78	1.82	1.92
		X	-	-	-	1.81	1.89	1.95
		XII	-	-	-	1.88	1.96	2.02
$_{56}Ba^{2+}$	$5p^6$	VI	1.43	1.34	1.35	1.36	1.44	1.50
		VIII	-	-	-	1.42	1.50	1.56
		IX	-	-	-	1.47	-	1.61
		X	-	-	-	1.52	1.60	1.66
		XII	-	-	-	1.60	1.68	1.74
$_{57}La^{3+}$	$4d^{10}$	VI	1.22	1.14	1.15	1.06	1.13	1.20
		VII	-	-	-	1.10	-	1.24
		VIII	-	-	-	1.18	1.26	1.32
		IX	-	-	-	1.20	-	1.34
		X	-	-	-	1.28	1.36	1.42
		XII	-	-	-	1.32	1.40	1.46

Ion	E.C.	C.N.	G	A	P	S-P	W-M	F-T
$_{67}Ho^{3+}$	$4f^{10}$	VI	-	.91	.97	.89	-	1.03
		VIII	-	-	-	1.02	-	1.16
$_{68}Er^{3+}$	$4f^{11}$	VI	-	.89	.96	.88	-	1.02
		VIII	-	-	-	1.00	-	1.14
$_{69}Tm^{3+}$	$4f^{12}$	VI	-	.87	.95	.87	-	1.01
		VIII	-	-	-	.99	-	1.13
$_{70}Yb^{3+}$	$4f^{13}$	VI	-	.86	.94	.86	-	1.00
		VIII	-	-	-	.98	-	1.12
$_{71}Lu^{3+}$	$4f^{14}$	VI	.99	.85	.93	.85	-	.99
		VIII	-	-	-	.97	-	1.11
$_{72}Hf^{4+}$	$4f^{14}$	VI	.85	.78	.81	.71	-	.85
		VIII	-	-	-	.83	-	.97
$_{73}Ta^{5+}$	$5p^6$	VI	.69	.68	-	.64	.72	.78
$_{74}W^{4+}$	$5d^2$	VI	.68	.70	-	.65	.73	.79
W^{6+}	$5p^6$	IV	-	-	-	.41	.50	.55
		VI	-	.62	-	.58	.68	.72
$_{75}Re^{4+}$	$5d^3$	VI	-	.72	-	.63	-	.77
$_{76}Os^{4+}$	$5d^4$	VI	.67	.69	-	.63	-	.77
$_{77}Ir^{4+}$	$5d^5$	VI	.66	.68	-	.63	-	.77
$_{78}Pt^{4+}$	$5d^6$	VI	-	.65	-	.63	-	.77
$_{79}Au^+$	$5d^{10}$	VI	-	1.37	1.37	-	-	-
$_{80}Hg^{2+}$	$5d^{10}$	VI	-	1.10	1.10	1.02	1.10	1.16
		VIII	-	-	-	1.14	1.22	1.28
$_{81}Tl^+$	$6s^2$	VI	1.49	1.47	1.40	1.50	-	1.64
Tl^{3+}	$5d^{10}$	VI	1.05	.95	-	.85	-	1.02
		VIII	-	-	-	1.00	-	1.14

Ion	Config.	CN	G	A	P	S-P	W-M	F-T
$_{58}Ce^{3+}$	$6s^1$	VI	1.18	1.07	1.11	1.03	1.09	1.17
		VIII	–	–	–	1.14	1.22	1.29
		XII	–	–	–	1.28	–	1.43
Ce^{4+}	$5p^6$	VI	1.02	.94	1.01	.80	.88	.94
		VIII	–	–	.97	1.05	1.11	
$_{59}Pr^{3+}$	$4f^2$	VIII	1.16	–	–	1.14	–	1.28
$_{60}Nd^{3+}$	$4f^3$	VIII	1.15	1.04	1.08	.995	–	1.135
$_{61}Pm^{3+}$	$4f^4$	VI	–	1.06	.98	.96	–	1.12
$_{62}Sm^{3+}$	$4f^5$	VI	–	1.00	1.04	.96	–	1.10
		VIII	–	–	1.09	–	1.23	
$_{63}Eu^{2+}$	$4f^7$	VIII	1.25	–	1.12	1.25	–	1.39
Eu^{3+}	$4f^6$	VI	–	.98	1.03	.95	–	1.09
		VIII	–	–	1.07	–	1.21	
$_{64}Gd^{3+}$	$4f^7$	VI	–	.97	1.02	.94	–	1.08
		VIII	–	–	1.06	–	1.20	
$_{65}Tb^{3+}$	$4f^8$	VI	–	.93	1.00	.92	–	1.06
		VIII	–	–	1.04	–	1.18	
$_{66}Dy^{3+}$	$4f^9$	VI	–	.92	.99	.91	–	1.05
		VIII	–	–	1.03	–	1.17	
$_{82}Pb^{2+}$	$6s^2$	VI	1.32	1.20	1.20	1.18	1.26	1.32
		VIII	–	–	–	1.29	1.37	1.45
		IX	–	–	–	1.33	–	1.47
		XII	–	–	–	1.49	1.57	1.63
Pb^{4+}	$5d^{10}$	VI	.84	.84	.84	.775	–	.945
		VIII	–	–	–	.94	–	1.08
$_{83}Bi^{3+}$	$6s^2$	V	1.1	–	–	.99	–	1.13
		VI	–	.96	–	1.02	–	1.16
		VIII	–	–	–	1.11	–	1.25
Bi^{5+}	$5d^{10}$	VI	–	.74	.74	–	–	–
$_{89}Ac^{3+}$	$6p^6$	VI	–	–	1.18	–	–	–
$_{90}Th^{4+}$	$6p^6$	VI	1.10	1.02	1.14	1.00	1.08	1.14
		VIII	–	–	–	1.06	1.12	1.20
$_{92}U^{4+}$	$5f^2$	VI	1.05	.97	.97	1.00	1.08	1.14
		VIII	–	–	–	1.05	–	1.19
U^{6+}	$6p^6$	II	–	.80	–	.45	–	.59
		VI	–	–	–	.75	.81	.89

Sources:

G: V. M. Goldschmidt (1926). Skrif.Norske Videnskaps-Akademie, Oslo, I, Mat. Naturviske. (1945) *Soil Sci.* 60, No. 1, July.

A: L. H. Ahrens (1952). *Geochim. et Cosmochim. Acta*, 2: 158.

P: Linus Pauling (1960). *The nature of the chemical bond*, 3d ed. Cornell Univ. Press, Ithaca, N.Y.

S–P: R. D. Shannon and C. T. Prewitt (1969). *Acta Crystallogr. Sect. B* 25: 925.

W–M: E. J. W. Whittaker and R. Muntus (1970) *Geochim. et Cosmochim. Acta*, 34: 952–3.

F–T: F. Fumi and M. T. Tosi (1964). *J. Phys. Chem. Solids*, 25: 31.

Their list of radii is extensive and includes values for each cation and anion in different coordination sites. Although Goldschmidt and Pauling were aware of the variation of ionic radius with change in CN, and provided correction factors for increasing or decreasing values of radii with corresponding changes in CN, they did not suggest variations in the radii of the anions with a change in cation CN. Shannon and Prewitt (as do Fumi and Tosi) provide different values for the O^{2-} and F^- anions when these are coordinated by II, III, IV, VI, or VIII cations. Implied, but not stated by these authors, is the presence of oxygen anions of different radius in the same mineral, as in beryl, which contains O^{2-} anions coordinated by both II and III cations. Shannon and Prewitt do give values that appear to be applicable to most minerals, but, because they include the effects of covalency, they also mask them. Normally, departure from ideal ionic sums of radii is used as a measure of covalency in minerals.

Still another set of empirically derived radii was presented by Whittaker and Muntus in 1970. They also provide different radii for cations and anions in different coordinations (see Table 4.2). The values of these investigators appear to lie between those of Shannon and Prewitt and of Fumi and Tosi.

In view of the large differences in ionic radii carefully derived by different investigators, it is surprising that the ionic model has been so successful in evaluation of crystal structure. The reason for such success lies in the fact that ionic and covalent models may yield the same result. If Pauling's concepts of electronegativity differences versus degree of ionic and covalent bonding, and resonance between ionic and covalent structures, are correct, then neither the ionic nor the covalent model per se accurately represents the chemical bonding in most minerals. An example of the two models, ionic and covalent, leading to similar results is afforded by the common mineral, sphalerite, ZnS, for which the interatomic distance is 2.35 Å (see Fig. 3.4). Radii are as follows:

Covalent sp^3 tetrahedral model	(Å)	Ionic model	(Å)
Zn_{sp^3}	1.31	Zn^{2+}	0.60
S_{sp^3}	1.04	S^{2-}	1.84
Sum of radii	2.35		2.44
	$R = r_{zn^{2+}}/r_{S^{2-}} = 0.326 = CN\ IV$		

Thus, both the extreme covalent sp^3 hybrid model and the extreme ionic model place four atoms or ions of S around Zn, and vice versa. According to Pauling's ideas, neither is correct. Pauling's more recent *electroneutrality concept* states that no atom ever carries a positive charge greater than unity; any greater amount of charge becomes neutralized by transfer of anionic electrons to the cation. This is a good place to close our discussion of ionic radii.

Summary

The covalent bond is directional or homopolar. The ionic bond is nondirectional or heteropolar.

The interatomic distance between atoms is governed by the balance of their attractive and repulsive forces.

The radius ratio (cation radius/anion radius) governs the packing of atoms around each other, or coordination.

Electrostatic bond strength is cation charge/cation CN, and is the fractional contribution of each cation to the charge on each anion it surrounds.

Bonds are never entirely ionic; percent ionic bonding may be estimated from differences in electronegativity and ionization potential of atoms in question. Alkali fluorides are the most ionic compounds.

The Si—O bond is 51% ionic and thus resonates as a mixture of ionic and covalent bonding.

The ionic bond requires spherical electron density: thus B-group elements with p valence electrons will form covalent complex anions.

Pauling's five rules codify the behavior of ionic compounds.

Rule 1. The number of anions surrounding a cation is a function of their radius ratio.

Rule 2. The number of coordination polyhedra sharing a common anion corner is limited by the EBS (cation charge/cation CN).

Rule 3. Sharing of corners, edges, and faces of coordination polyhedra leads to their distortion, and decreases stability, because of the electrostatic repulsion of cations.

Rule 4. Polyhedra with small cations of high charge tend not to share anions.

Rule 5. A large number of different kinds of coordination polyhedra in a given mineral tends to decrease its stability. There are many exceptions to this rule.

The closeness of packing of spheres of equal radius is the same for both hexagonal and cubic close-packed arrays. In both HCP and CCP, each anion is the common corner of eight tetrahedra and six octahedra. A maximum of six of these fourteen cation coordination sites is occupied in minerals, with the cations placed in geometric arrays, equidistant from one another. The radius ratio of rigid spheres governs the placement of small spheres in the interstices of large spheres. Minimum radius ratios are tetrahedral (CN IV, $R = 0.225$), octahedral (CN VI, $R = 0.414$), square antiprismatic (CN VIII, $R = 0.645$), cubic (CN VIII, $R = 0.732$), and cubo–octahedral or hexagonal–antiprismatic (CN XII, $R = 1.0$).

Ionic radii are ideal distributions of interatomic distances derived from X-ray analyses. They have been derived by different methods by

Wasastjerna, Goldschmidt, Pauling, Ahrens, Fumi and Tosi, Shannon and Prewitt, and Whittaker and Muntus. All these ionic radii are tabulated and compared.

Bibliography

Ahrens, L. H. (1952). The use of ionization potentials. Part 1. Ionic radii of the elements. *Geochim. et Cosmochim. Acta,* 2: 155–69.

Brown, G. E., and Gibbs, G. V. (1969). Oxygen coordination and the Si—O bond. *Am. Mineral.,* 54: 1528–39.

Fumi, F. G., and Tosi, M. P. (1964). Ionic sizes and Born repulsive parameters in the NaCl-type alkali halides. I. The Huggins–Mayer and Pauling forms. *J. Phys. Chem. Solids,* 25: 31–43.

Goldschmidt, V. M. (1954). *Geochemistry.* Clarendon Press, Oxford.

Pauling, L. (1960). *The nature of the chemical bond,* 3rd ed. Cornell Univ. Press, Ithaca, N.Y., Chap. 13.

(1987). Determination of ionic radii from cation–anion distances in crystal structures: Discussion. *Am. Mineral, 72:* 1016.

Shannon, R. D., and Prewitt, C. T. (1969). Effective ionic radii in oxides and fluorides. *Acta Crystallogr. Sect. B,* 25: 925–45.

Tosi, M. P., and Fumi, F. G. (1964). Ionic sizes and Born repulsive parameters in the NaCl-type alkali halides. II. The generalized Huggins–Mayer form. *J. Phys. Chem. Solids,* 25: 45–52.

Wells, A. F. (1970). *Models in structural inorganic chemistry.* Oxford Univ. Press, New York.

Whittaker, E. J. W., and Muntus, R. (1970). Ionic radii for use in geochemistry. *Geochim. et Cosmochim. Acta,* 34: 952–3.

Wondratschek, Hans (1987. Determination of ionic radii from cation–anion distances in crystal structures. *Am. Mineral.,* 72: 82.

5 Pauling's second rule of electrostatic valency in ionic or coordination compounds

Whereas Pauling's first rule uses ionic radii of cation and anion and their radius ratio R to establish the number of anions that surround each cation (cation CN), it is Pauling's second rule that employs electrostatic valency or bond strength to limit the number of cations that surround each anion (anion CN). Implicit in the second rule is the requirement that electrical charge of anion and cation be neutralized locally by nearest neighbors. In minerals containing one species of anion, such as O^{2-} or F^-, it might be tacitly assumed that the cation array or cation environment around each anion is identical. In many minerals, this is true, and it may be seen that in the mineral rutile, TiO_2, each O^{2-} anion is surrounded by three Ti^{4+}/VI cations that contribute fractional electrostatic bond strengths of $4+/VI$, exactly satisfying locally the negative valency of each O^{2-} anion. Similarly, in the mineral fluorite, CaF_2, each F^- anion is surrounded by four $Ca^{2+}/VIII$ cations, each contributing a fractional electrostatic bond strength of $2+/VIII$ exactly satisfying locally the univalency of each F^- anion. We will follow the practice of using schematic *electrostatic bond strength (EBS) diagrams,* illustrated in Figure 5.1, to express electrostatic valency solutions, and to derive the structural formula for a given mineral.

We will classify rutile and fluorite into the category Pauling Rule 2A group, representing those minerals that contain one unique anion and, hence, one unique EBS solution; all anions are thus equivalent. Other minerals may contain two or more anions with different cation environment, or they may contain two different anions of different valency, such as O^{2-} and F^-. In such minerals, each different anion will require a different electrostatic valency solution. We will classify these in the category Pauling Rule 2B group.

In a third category, the Pauling Rule 2C group, we will include those minerals in which the same anion (e.g., O^{2-}) is surrounded by two or more different cation arrays, none of which satisfies Pauling's second rule, because each anion is either undercharged or overcharged. However, in these minerals electrostatic neutrality is maintained by being spread out over a larger domain. For example, the sum of bond strengths reaching three different O^{2-} will total 6+, although each will individually total more or less than the required 2+. Thus, in the pyroxene mineral group, diopside, $CaMgSi_2O_6$, contains three different oxygen anions; O-1, O-2, and O-3, which are surrounded, respectively, by cation arrays contributing EBS summations of +1.92, +1.58, and +2.50, for a total of 6+. In such minerals it has been observed from X-ray studies that the cations Ca^{2+}, Mg^{2+}, and Si^{4+} do not lie in the centers of their respective coordination polyhedra, because they move away from overcharged anions and closer to undercharged anions. This led the eminent crystallographer Zachariasen (1963) to postulate that EBS is inversely proportional to bond length. We may refer to this idea as Zachariasen's elaboration of Pauling's second rule. It cannot be applied without detailed values of interatomic distances obtainable only from X-ray analysis of crystals. We will now consider several examples of each of these categories of Pauling electrostatic valency types.

Pauling Rule 2A group

We have already cited rutile and fluorite as examples of minerals containing one unique anion. A more complex representative of this category is

Figure 5.1. Electrostatic bond strength diagrams for minerals in which all oxygens have equivalent near neighbor cation arrays: (A) rutile; (B) fluorite; (C) garnet; (D) zircon; (E) xenotime. All oxygen anions in structure are identical.

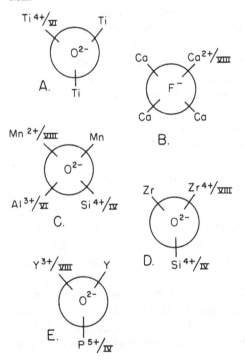

the garnet mineral group. The mineral spessartine is a garnet of the composition $Mn_3Al_2[SiO_4]_3$, in which cation coordination numbers, predicted from radius ratios, place each Mn^{2+} ion in CN VIII, each Al^{3+} ion in CN VI, and each Si^{4+} ion in CN IV. Electrostatic valency considerations, now confirmed by X-ray location of the ions, require that each O^{2-} anion be surrounded by two Mn^{2+} ions contributing $2+/VIII$, one Al^{3+} ion contributing $3+/VI$, and one Si^{4+} ion contributing $4+/IV$, as illustrated in Figure 5.1 (also see Figs. 12.7, 12.8).

At first, it might appear that this arrangement would not be compatible with the garnet formula, but, because the positive charge on each cation is divided equally among all neighboring anions, the garnet formula may be derived from the bond strength diagram as follows:

Mn^{2+}	CN VIII	$\frac{1}{8}$ of each 2 Mn reaches an O^{2-} ion
Al^{3+}	CN VI	$\frac{1}{6}$ of each Al reaches an O^{2-} ion
Si^{4+}	CN IV	$\frac{1}{4}$ of each Si reaches an O^{2-} ion

Thus, it follows that

$$\left.\begin{array}{l} \frac{1}{8}\, Mn^{2+} \\ \frac{1}{8}\, Mn^{2+} \\ \frac{1}{6}\, Al^{3+} \\ \frac{1}{4}\, Si^{4+} \\ 1\, O \end{array}\right\} \times 12 \qquad \begin{array}{l} Mn_3 \\ \\ Al_2 \\ Si_3 \\ O_{12} \end{array}$$

Here, it is important to know that whenever the grouping $[SiO_4]$ appears in a mineral formula, the $[SiO_4]$ tetrahedra are independent of one another and there will be but one unique O^{2-} anion in the structure. Thus, in minerals such as olivine, $(Mg,Fe)_2[SiO_4]$ and zircon, $Zr[SiO_4]$, each O^{2-} anion will be in an identical cation environment. In olivine, each and every O^{2-} anion will be surrounded by one Si, $4+/IV$, and three (Mg,Fe), $2+/VI$, for a total EBS of $2+$. In zircon, each O^{2-} anion will be surrounded by one Si, $4+/IV$, and two Zr, $4+/VIII$, ions for a total EBS of $2+$ (Fig. 5.1). In xenotime, YPO_4, isostructural with zircon, each O^{2-} anion is bonded by one P^{5+}/IV and two $Y^{3+}/VIII$ ions in a zircon-type lattice array (Figs. 5.1, 12.5, 12.6).

Pauling Rule 2B group

Whenever SiO_4 tetrahedra share corners with one another, it is no longer possible for all O^{2-} anions to be equivalent. An oxygen anion common to two tetrahedral corners will be in contact with two Si^{4+} ions, and $4+/IV \times 2$ will completely saturate the valency of the common O^{2-} anion. An oxygen anion that links one or more corners of two polyhedra is called a *bridging oxygen*. The bridging oxygen will thus have a different cation environment from nonbridging oxygens. A simple example is provided by the mineral thortveitite, $Sc_2[Si_2O_7]$. Here, two SiO_4 tetrahedra are joined by a bridging oxygen to produce the dumbbell-shaped Si_2O_7 group. The bridging oxygen will be in contact with two Si^{4+} ions, and the six remaining oxygens, nonbridging, will each be in contact with only one Si^{4+} ion and two Sc^{3+} ions, the latter in CN VI. The bond strength diagram shown in Figure 5.2 illustrates the presence of two different O^{2-} anions in the structure, implicit in the formula $Sc_2[Si_2O_7]$. (Also see Fig. 12.2.)

Another good example of this category is provided by the mineral beryl, $Al_2Be_3[Si_6O_{18}]$. Here, each SiO_4 tetrahedron shares two of its corner oxygens with adjoining tetrahedra to build hexagonal Si_6O_{18} rings. The six O^{2-} anions forming the center of each ring will be in contact with two Si^{4+} cations, completely neutralizing their charge, and thus may not be in contact with either Al^{3+} or Be^{2+}. The remaining twelve O^{2-} anions will each be in contact with one Si, 4+/IV, one Be, 2+/IV, and one Al, 3+/VI, for a total of 2+. The bond strength diagram for beryl is shown in Figure 5.2. (Also see Fig. 14.2.)

In thortveitite, Si_2O_7 dumbbell groups are linked together by Sc^{3+} ions; in beryl, Si_6O_{18} hexagonal rings are joined together by Al^{3+} and Be^{2+} ions. Each mineral contains nonequivalent O^{2-} ions.

In the sheet silicate or phyllosilicate minerals, each SiO_4 tetrahedron shares three of its four corners with adjoining tetrahedra to form infinite two-dimensional Si_4O_{10} sheets. It will be apparent that the three common tetrahedral corner oxygens will each be in contact with two Si^{4+} ions that will completely neutralize their charge. The fourth, unshared, oxygen will be in contact with only one Si^{4+} ion, and it may contact other cations.

In the mineral talc, $Mg_3[Si_4O_{10}](OH)_2$, the unshared O^{2-} anion will be in contact with three Mg cations, 2+/VI, in addition to one Si ion, 4+/IV, for a total of 2+. The hydroxyl ion, $(OH)^-$, will contact three Mg ions but may not be in contact with an Si ion. Thus, in talc, we have three different anions,

Figure 5.2. Electrostatic bond strength distribution in minerals containing two nonequivalent oxygens, with nonequivalent near neighbor cation arrays. (A) thortveitite; (B) beryl.

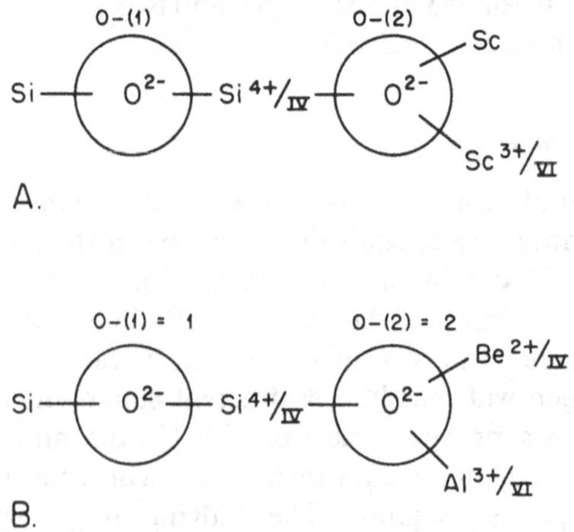

each having its negative valency satisfied in different ways by nearest cation neighbors. We may organize these as a sandwich, with Mg cations lying in between and joining opposed tetrahedral $Si_2O_5(OH)_2$ sheets, as schematically illustrated in Figure 5.3 (also see Fig. 15.2). We have three basal oxygens, O-(1), one apical oxygen, O-(2), in each tetrahedron, and one (OH) ion outside of each tetrahedron. The bond strength diagram of talc is shown in Figure 5.3.

Some minerals from Group 2B incorporate anions of two different valencies into their structure, necessitating two different cation environments. A good example is the rare earth carbonate mineral bastnaesite, $CeF[CO_3]$, containing the anions O^{2-} and F^-. In this mineral, the relatively large cerium ion Ce^{3+} has radius of 1.14 Å for CN VIII and 1.29 Å for CN XII, yielding radius ratios with the O^{2-} anion of 0.838 and 0.948, respectively, which are compatible with either CN VIII or CN XII. The EBS solution, however, suggests that the CN of Ce^{3+} is actually IX rather than VIII or XII. X-ray analysis shows that each Ce^{3+} ion is coplanar with three F^- anions and in contact with these, plus three O^{2-} above and another three O^{2-} below in a geometrically irregular CN IX site. Each F^- anion is surrounded by three Ce^{3+} near neighbors; each O^{2-} anion is in turn surrounded by two Ce^{3+} and one C^{4+} ion of a $[CO_3]$ group. Each carbon atom is actually covalently bonded to a distorted planar triad of O^{2-} anions in a $[CO_3]^{2-}$ anion complex. Thus, each such $[CO_3]^{2-}$ planar anion is surrounded

Figure 5.3. Electrostatic bond strength distribution in minerals containing three different anions, as in talc, O-(1), O-(2), and (OH) = O-(3).

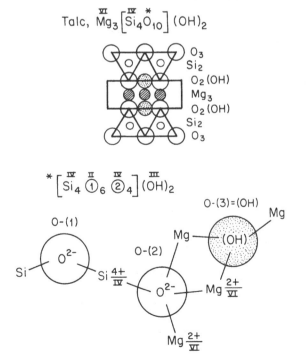

by six trivalent cerium ions to yield an electrostatic bond summation of Ce^{3+}/
$IX = 18/9 = +2$, and three such Ce bonds reach each F^- anion for a summation of $+1$. The EBS diagram is illustrated in Figure 5.4A (also see Figs. 19.7, 19.8).

Still another example is provided by the cubic mineral sulfohalite, $Na_6[SO_4]_2FCl$, which may be alternatively written as $2Na_2[SO_4] \cdot NaF \cdot NaCl$. This mineral has three different anions: divalent oxygen, univalent fluorine, and univalent chlorine. The elegant cubic structure is built of tetrahedra of oxygen anions around each sulfur ion, $[SO_4]^{2-}$, and octahedra of four oxygen anions plus one fluorine anion and one chlorine anion. The four O^{2-} anions around each Na^+ cation are corner oxygens of four different $[SO_4]^{2-}$ tetrahedra, and the F^- and Cl^- anions are centers of octahedral groupings of six Na^+ cations. The coordination numbers of each ion are summarized below, and the EBS diagram is given in Figure 5.4B. The crystal structure is further illustrated in Figure 5.5 to clarify the polymerization of polyhedra.

Sulfohalite $Na_6^{VI}[SO_4^{XII}]_2F^{VI}Cl^{VI}$ or

$2Na_2^{VI}[S^{IV}O_4^{IV}] \cdot Na^{VI}F^{VI} \cdot Na^{VI}Cl^{VI}$

Each S^{6+} is coordinated by IV O^{2-}
Each Na^+ is coordinated by IV O^{2-}, 1 F^-, 1 $Cl^- = CN$ VI
Each F^- is coordinated by VI Na
Each Cl^- is coordinated by VI Na^+

Figure 5.4. Electrostatic bond strength distribution in minerals with different species of anions, O, F, and Cl: (A) bastnaesite; (B) sulfohalite.

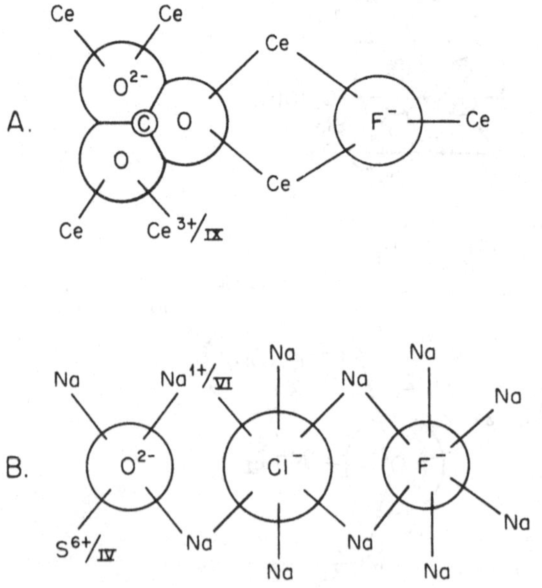

Each O^{2-} is coordinated by III Na^+, 1 S^{6+} = CN IV
Each $[SO_4]^{2-}$ is coordinated by XII Na^+

Pauling Rule 2C group

In many minerals, the nearest neighbor cation arrays around each anion may differ, leading to EBS summations that depart widely from the ideal of $+2.00$ for oxygen or $+1.00$ for fluorine anions. Neutrality, then, is not satisfied locally, but must be spread out over a larger domain in the crystal, theoretically resulting in a reduction in stability. Consider the example of diopside, $CaMg[Si_2O_6]$, where three different cation arrays surround oxygens labeled 1, 2, and 3, leading to idealized EBS strength summations of 1.916 around 1, 1.583 around 2, and 2.500 around 3 (Fig. 5.6). In place of the simple formula $Ca^{VIII}Mg^{VI}[Si_2^{IV}O_6]$, we may write $Ca^{VIII}Mg^{VI}[Si_2^{IV}①_2^{IV}②_2^{III}③_2^{IV}]$ to emphasize the presence of the three unequal or different oxygens. From Figure 5.6 and Table 5.1, we see that although none of the three oxygens strictly satisfies Pauling's Rule 2, the sum of the EBS reaching all three oxygens is $+6.00$, and electric neutrality is assumed to be spread over the three oxygens rather than around each. From the data of Table 5.1, we see that Si, Mg, and Ca

Figure 5.5. Packing model of part of the sulfohalite structure. Large spheres represent O^{2-}, F^-, and Cl^- anions. Corks represent Na^+ ions, and small white spheres, S^6 ions in $[SO_4]^{2-}$ groups.

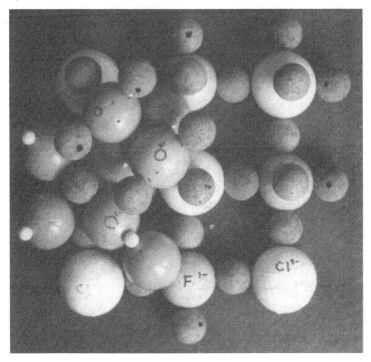

Table 5.1. *Measured and ideal interatomic distances in diopside, and electrostatic bond strength solutions derived from these*

Anion–cation arrays	A—B (Å) obs.	EBS Zachariasen	EBS Pauling	A—X (Å) Ideal
① Si	1.603	1.12	1.00	1.613
① Mg	2.175	0.21	0.333	2.080
① Mg	2.061	0.40	0.333	2.080
① Ca	2.374	0.28	0.250	2.480
		2.01	1.916	
② Si	1.591	1.26	1.00	1.613
② Mg	2.053	0.40	0.333	2.080
② Ca	2.332	0.34	0.250	2.480
		2.00	1.583	
③ Si	1.675	0.82	1.00	1.613
③ Si	1.690	0.80	1.00	1.613
③ Ca	2.606	0.22	0.250	2.480
③ Ca	2.746	0.16	0.250	2.480
		2.00	2.500	

Note: ① = O-(1) = oxygen designation number.
Source: Clark, Appleman, and Papike (1969).

Figure 5.6. Cation–anion shifts in *C2/c* pyroxene, diopside, resulting from excess electric charge on anions O-(3) and insufficient charge on anions O-(1) and O-(2). See also Table 5.1 for bond strength distribution on all three anions.

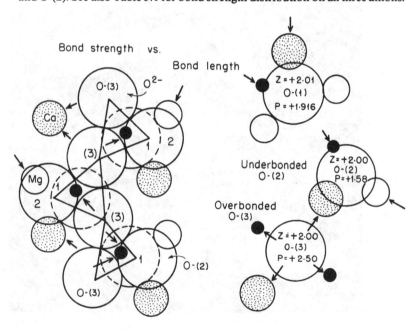

cations are not centrosymmetrically disposed in their oxygen polyhedra, and that such polyhedra (tetrahedron, octahedron, and square antiprism) are considerably distorted from ideal geometry (see Fig. 8.1); we shall return to this subject of polyhedral distortion in Chapter 8.

Zachariasen's elaboration of Pauling's Rule 2

Zachariasen (1963) suggested that EBS is inversely proportional to A—X (Å) bond lengths as accurately measured from single crystal X-ray analyses of borate minerals. Thus an oxygen with a classical Pauling bond strength of less than +2.00 will increase its effective bond strength by drawing closer to neighboring cations than is normal or ideal: bond lengths shorten. Conversely, where the bond strength of an anion exceeds +2.00, bond strength is decreased as bonds to neighboring cations lengthen. Bond strengths empirically assigned on such a basis of departure from ideal A—X distances lead to localized charge balances on each anion, as shown in Figure 5.6 and Table 5.1 for diopside.

Summary

Pauling's second rule employs EBS to limit the number of cations around each anion and the number of polyhedra with a common corner.

Group 2A includes minerals with one species of anion, each with the same environment.

Group 2B includes minerals whose anions have a different environment or different valency.

Group 2C includes minerals each of whose anions is surrounded by a different cation array, the sum of whose bond strengths satisfies electrostatic neutrality, although no one of them is electrostatically neutral.

Each of these groups is illustrated by several mineral examples.

Zachariasen's elaboration of this rule suggests that EBS is inversely proportional to bond length; that is, metal–anion distances will be shorter than ideal for electrically underbonded anions. Here is a prime cause of polyhedral distortion.

Bibliography

Clark, J. R., Appleman, D. E., and Papike, J. J. (1969). Crystal chemical characterization of clinopyroxenes based on eight new structure refinements. *Mineral. Soc. Am. Spec. Pap.*, 2: 31–50.

Pabst, A. (1934). The crystal structure of sulfohalite. *Z. Krist.*, 89: 514.

Pauling, L. (1960). *The nature of the chemical bond,* 3d ed. Cornell Univ. Press, Ithaca, N.Y.

Watanabé, T. (1934). The crystal structure of sulfohalite. *Proc. Imp. Acad. (Tokyo),* 10: 575.

Zachariasen, W. H. (1963). The crystal structure of monoclinic metaboric acid. *Acta Crystallogr., Pt. 5* 16: 385–92.

6 External (nontranslational) and internal (translational) symmetry

Crystals may be regarded as solids enclosing a three-dimensional lattice or array of points that translates its pattern geometrically in space. There are but five two-dimensional lattices from which all minerals are built. These outline points in space as a

1. Square
2. Rectangle
3. Diamond with angles between the edges not 120° or 60°
4. Rhombus with angles of 120° and 60°
5. Parallelogram with angles neither 60°, 90°, nor 120°, and with sides unequal

When these two-dimensional lattices are combined with a second such perpendicularly oriented two-dimensional lattice, a three-dimensional array of points, the *Bravais space lattice,* is built.

It is geometrically possible to construct only fourteen such unique space lattices (Fig. 6.1). Of these fourteen lattices, the seven labeled *P*, for primitive (Fig. 6.1) represent the seven crystal systems into which all minerals and regular solids may be classified and readily recognized (Fig. 6.2).

Although it is possible to outline or enclose, from an extended array of lattice points, more than one such set that repeats itself (translates itself), we choose the smallest coplanar array of points that translates itself in three-dimensional space to define the unit cell. For halite, NaCl, Cl atoms or ions on eight corners of the cell and on all six face centers, and Na atoms or ions on all twelve edges and in the center, outline the unit cube marked D on Figure 6.3. Cubelet A (Fig. 6.3), which is smaller, does not repeat itself in the space array of points. Cubelet B, though smaller than D, repeats itself in our two-dimensional array, but will not repeat itself in vertical translation because Na ions will be positioned directly over the Cl ions shown. Cube C is too large. Hence cube D becomes the only choice for the familiar unit cell of halite (also see Fig. 10.1).

If we closely stack together unit blocks or cells of a mineral, it will be apparent that atoms located on corners are common to eight unit cells, those on edges to four, those on faces to two, and those inside to one. Thus, for halite the unit cube of Figure 6.3D contains:

$$\text{Cl} \longrightarrow 8_{cor} \times \tfrac{1}{8} = 1 \quad \text{and} \quad 6_{faces} \times \tfrac{1}{2} = 3; \qquad \text{sum} = 4\,\text{Cl}$$
$$\text{Na} \longrightarrow 12_{edge} \times \tfrac{1}{4} = 3 \quad \text{and} \quad 1_{ctr} = 1; \qquad \text{sum} = 4\,\text{Na}$$

Figure 6.1. The fourteen Bravais space lattices. Reprinted with permission from M. J Buerger, *Elementary crystallography*, John Wiley & Sons, 1956.

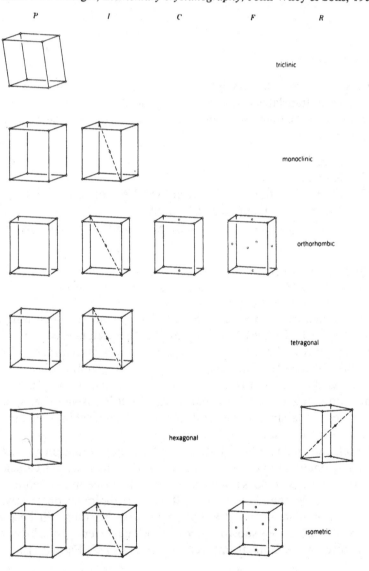

Figure 6.2. Symmetry axes in crystal drawings representing the seven primitive crystal systems. Symbols as in Figure 6.5.

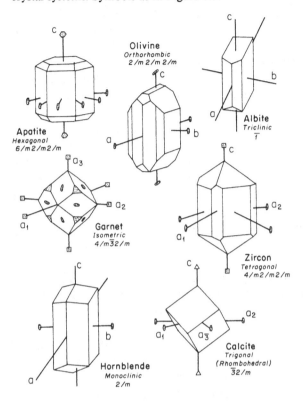

Figure 6.3 Two-dimensional repeat cells in packing model of halite, NaCl. The true unit cell with smallest repeat unit of all atoms is labeled D.

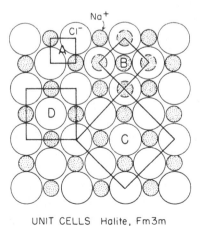

UNIT CELLS   Halite, Fm3m

A very small crystal of halite balanced on the tip of one's finger represents a cube made up of many smaller cubelets, the smallest being the unit cube bounded by twelve edges, a (or a_0), of length 5.46 Å (5.46×10^{-8} cm). Without X-ray data, however, our small halite crystal represents a cube that reveals nothing about the internal distribution of the Na and Cl ions of which it is built, and does not allow us to identify the space lattice.

The geometry or external symmetry of the cube (and of all cubes) does display four threefold axes of rotational symmetry passing through cube corners, three fourfold axes passing through the six cube faces, each perpendicular to a mirror plane of symmetry that bisects each cube face parallel to its edges, six twofold axes of symmetry that pass through the opposite twelve edges and each perpendicular to planes of mirror symmetry that pass through each cube diagonal. Because opposite faces, edges, and corners are identical, the crystal has a center of symmetry. Now, thirteen axes + nine mirror planes + the center add up to twenty-three elements of nontranslational (external) symmetry. Thus it is possible to classify the cube into a *crystal class* (point group) of the highest possible symmetry, coded $4/m\,\bar{3}\,2/m$ of Table 6.1.

There are thirty-two of these crystal classes (Table 6.1), each displaying lower external symmetry in passing from the isometric to the triclinic crystal systems. Class $4/m\,\bar{3}\,2/m$ is one of five classes that have isometric symmetry; it is also one of thirty-six space groups that have isometric internal symmetry (Table 6.2).

To identify which of the fourteen space lattices and which of the 230 space groups are represented, we must use X-ray analyses to determine internal symmetry. Translational symmetry elements are screw axes and glide planes that translate internal lattice points some regular increment of a unit cell dimension or length when combined with rotation (Figs. 6.4, 6.5; see also Figs. 3.3, 10.1). Atomic positions are located by single crystal X-ray analyses (Weissenberg patterns), whereas different planes of atoms that repeat themselves by three-dimensional translation are identified by X-ray powder diffraction analysis (Debye–Scherrer patterns) of a minute amount of finely powdered material. Usually, X-ray powder diffraction analysis serves to identify the crystal, the space lattice, and the space group. A single crystal X-ray analysis may then be made to locate specific atoms within the lattice, based on the capacity of different atomic species to scatter X-radiation of a particular wavelength. Atoms distant from one another in the periodic table are readily identified. Atoms with similar atomic number, or next to one another, such as $_{13}$Al and $_{14}$Si, will scatter X-radiation to an almost identical degree and render their discrimination on this basis impossible. Thus, to determine whether a particular tetrahedron in feldspars is occupied by Al or Si, we must measure the T—O (tetrahedral cation–oxygen) distance and the O—O distance to decide that the larger tetrahedral site contains the Al and the smaller contains the Si.

X-ray powder diffraction patterns are made on film in Debye–Scherrer cameras or on automatic recording charts with X-ray diffractometers. These pat-

Table 6.1. *Symmetry components of the thirty-two ...*

Crystal system	Axes 2	Axes 3	Axes 4	Axes 6	Planes m	Center c	Σ	H–M class	
Triclinic $a,b,c \neq a \wedge b = r,\ b \wedge c = \alpha,\ a \wedge c = \beta \neq 90°$							0	1	1
						1	1	$\bar{1}$	2
Monoclinic $a,b,c \neq a \wedge c = \beta \neq 90°$	1						1	2	3
					1		1	m	4
	1				1	1	3	2/m	5
Orthorhombic $a,b,c \neq a \perp b \perp c \perp a$	3						3	2 2 2	6
	1				2		3	m m 2	7
	3				3	1	7	2/m 2/m 2/m	8
Trigonal $a_1 a_2 a_3 = \perp c \neq$; $a_1 \wedge a_2 \wedge a_3 = 120°$		1					1	3	9
		1				1	2	$\bar{3}$	10
	3	1					4	3 2	11
		1			3		4	3 m	12
	3	1			3	1	8	$\bar{3}$ 2/m	13
Tetragonal $a_1 = a_2 \perp c \neq$; $a_1 \wedge a_2 \wedge c = 90°$			1				1	4	14
			1				1	$\bar{4}$	15
			1		1	1	3	4/m	16
	4		1				5	4 2 2	17
			1		4		5	4 m m	18
	2		1		2		5	$\bar{4}$ 2 m	19
	4		1		5	1	11	4/m 2/m 2/m	20
Hexagonal $a_1 a_2 a_3 = \perp c \neq$; $a_1 \wedge a_2 \wedge a_3 = 120°$				1			1	6	21
				1			2	$\bar{6}$	22
				1	1	1	3	6/m	23
	6			1			7	6 2 2	24
				1	6		7	6 m m	25
	3			1	3		7	$\bar{6}$ m 2	26
	6			1	7	1	15	6/m 2/m 2/m	27
Isometric (cubic) $a_1 = a_2 = a_3$, all 90°	3	4					7	2 3	28
	3	4			3	1	11	2/m$\bar{3}$	29
	6	4	3				13	4 3 2	30
		4	3		6		13	$\bar{4}$ 3 m	31
	6	4	3		9	1	23	4/m $\bar{3}$ 2/m	32

terns show many lines of differing intensity and location or interplanar spacing, which result from the in-phase scattering of X-radiation by each atom in a plane of atoms (although constructive waves enter and leave at the same angle to a crystal plane, and without a change in wavelength, diffraction by atoms in a plane is not directly analogous to optical reflection).

The location of X-ray diffraction lines depends on the wavelength of radiation used (Cu and Fe anodes are the most common), the angle of the collimated X-ray beam with a plane of atoms, and the distance (interplanar spacing, d [Å]) between planes according to Bragg's law of diffraction, $n\lambda = 2d \sin \theta$, where n is an integral number of wavelengths of the X-radiation used, d is the interplanar spacing, and θ is the angle of incidence (Fig. 6.6). The intensity of the particular lines in a pattern is largely a function of the atomic scattering factor (heavy atoms scatter X-radiation more efficiently) and, to some degree, of the angle of incidence and the type of X-ray target (anode) used.

In many cases, space lattice and space group identifications depend as much on the absence of particular Bragg X-ray reflections (diffraction lines) as on the presence of others, because the introduction of a like plane of atoms, halfway between two identical planes of atoms, say at 4.00 Å, will cancel the 4.00 Å spacing and yield a 2.00 Å spacing. Thus, the 5.46 Å spacing of the unit cell edge of halite is canceled by the introduction of the identical plane of face centered atoms at 2.73 Å: a strong line at 2.73 Å thus represents half the distance of the unit cell edge and is given the Bragg diffraction line label d_{200} rather than d_{100}. The absence of the d_{100} reflection and other systematic absences from other planes in the crystal permit the identification of the space lattice as F (face-centered) for halite. Note that if the lattice had been P (primitive) with points only at the corners, both the d_{100} and d_{200} Bragg reflections would have

Table 6.2. *Crystal systems, space lattices, crystal classes, space groups for all minerals and synthetic crystals*

Systems	Lattices & symbols	No. of crystal classes	No. of space groups
Isometric	3, P, I, F	5	36
Tetragonal	2 $P (=C)$, $I (=F)$	7	68
Hexagonal	1 $P (=C)$	7	27
Trigonal	1 $P (=R)$	5	25
Orthorhombic	4 P, C, I, F	3	59
Monoclinic	2 P, C	3	13
Triclinic	1 P	2	2
Total 7	14	32	230

appeared. The student will profit from learning, from published data, the patterns of Bragg reflections that are produced from different types of space lattices.

The description of a space lattice as primitive, *P*, body centered, *I*, or face centered, *F*, from a drawing or projection of atomic locations, will be obvious from Figure 6.1. Note that in the isometric system the centering of a point on

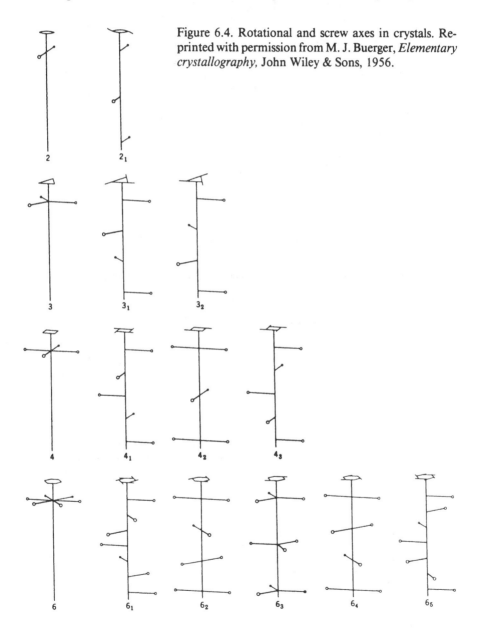

Figure 6.4. Rotational and screw axes in crystals. Reprinted with permission from M. J. Buerger, *Elementary crystallography*, John Wiley & Sons, 1956.

Figure 6.5. Symbols used for space group notations.

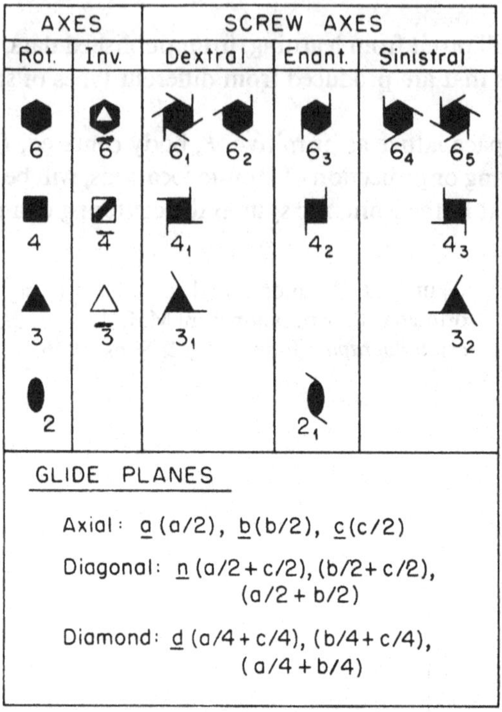

AXES		SCREW AXES		
Rot.	Inv.	Dextral	Enant.	Sinistral
6	$\bar{6}$	6_1 6_2	6_3	6_4 6_5
4	$\bar{4}$	4_1	4_2	4_3
3	$\bar{3}$	3_1		3_2
2			2_1	

GLIDE PLANES

Axial: $\underline{a}\,(a/2)$, $\underline{b}\,(b/2)$, $\underline{c}\,(c/2)$

Diagonal: $\underline{n}\,(a/2+c/2)$, $(b/2+c/2)$, $(a/2+b/2)$

Diamond: $\underline{d}\,(a/4+c/4)$, $(b/4+c/4)$, $(a/4+b/4)$

Figure 6.6. Graphic solution of Bragg's law of X-ray diffraction, $n\lambda = 2d \sin \theta$.

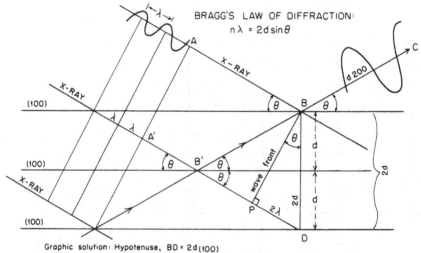

BRAGG'S LAW OF DIFFRACTION:
$n\lambda = 2d\sin\theta$

Graphic solution: Hypotenuse, BD = 2d(100)
 Adjacent side, BP = Wave front, opposite side, PD = 2λ

λ = Wavelength in Å of X-ray source
θ = Angle of incidence (glancing angle) with crystal planes
n = Integral number of wavelengths, λ

78

Table 6.3. *Space groups in which common minerals crystallize*

Triclinic Key: no planes of symmetry; only a center and a 1 (1/360°) axis = no symmetry	$P, C, 1$ or $\bar{1}$ $C\bar{1}$ $I\bar{1}$ $P\bar{1}$			Kaolinite, Microcline, Albite, Bytownite, Anorthite
Monoclinic Key: one twofold axis + 1 plane of symmetry, or 1 of either	$P, C, (A, B)$ $C2/c$ $P2_1/a$ $P2_1/m$ $P2_1/c$ $C2/m$	$b/(010)$		Augite, Diopside, Muscovite, Datolite, Epidote, Pigeonite, Sanidine, Phlogopite
Orthorhombic Key: 3 letters or numbers or their combination after the lattice symbol	P, C, I, F P P P P F A	(100) b n n b d m	(010) n n m c d m	(001) m m a a d a — Diaspore, Forsterite, Sillimanite, Andalusite, Anthophyllite, Hypersthene, Sulfur, Anhydrite
Tetragonal Key: 4 in first position (*No* 3-axis)	$P (=C), I (=F)$ $P4_2/$ $I4_1/$ $P\bar{4}2_1$	$c/(001)$ m a	(100) n m	(110) m d m — Rutile, Anatase, Zircon, Melilite
Hexagonal Key: 6 in first position	$P (=C)$ $P6_3/$ $P6_3/$ $P\bar{6}$	$c/(0001)$ m m	$(11\bar{2}0)$ m 2	$(10\bar{1}0)$ c c — Molybdenite, Graphite, Apatite, Bastnaesite, Benitoite
Trigonal Key: 3 in first position	$R (=P)$ $R\bar{3}$ $R3$ $R\bar{3}$	$c/(0001)$	$(11\bar{2}0)$ m	$(10\bar{1}0)$ c — Dolomite, Ilmenite, Tourmaline, Calcite, Corundum, Hematite

Table 6.3. *(cont.)*

Isometric	P, I, F	(001)	(111)	(110)	
Key: 3 in second position					
	F	m	3	m	⎰ Fluorite / Galena / Halite
	F	d	3	m	⎰ Spinel / Magnetite / Diamond
	P	n	3	m	Cuprite
	I	a	3	d	⎰ Analcime / Garnet
	P	a	3		Pyrite
	$P\bar{4}$		3	n	Sodalite
	$F\bar{4}$		3	m	Sphalerite

Table 6.4. *Minerals of diverse bond type crystallized in space group Pbnm and the related groups Pnma and Pcmn*

Pbnm

Goethite	$FeO(OH)$	Covalent–ionic
Forsterite	$Mg_2[SiO_4]$	Ionic–covalent
Ramsdellite	MnO_2	Ionic-metallic
Chrysoberyl	$Al_2[BeO_4]$	Ionic-covalent
Topaz	$Al_2[SiO_4](OH,F)_2$	Ionic-covalent
Danburite	$Ca[B_2Si_2O_8]$	Ionic-covalent
Stibnite	Sb_2S_3	Metallic-covalent
Diaspore	$AlO(OH)$	Ionic-covalent
Sillimanite	$AlOSi[AlO_4]$	Ionic-covalent
Humite	$Mg(OH,F)_2 \cdot 3Mg_2SiO_4$	Ionic-covalent
Sinhalite	$MgAl[BO_4]$	Ionic-covalent

Pnma

Anthophyllite	$(Mg,Fe)_7[Si_8O_{22}](OH)_2$	Ionic-covalent
Barite	$Ba[SO_4]$	Ionic-covalent
Chalcostibite	$CuSbS_2$	Metallic-covalent

Pcmn

Aragonite	$Ca[CO_3]$	Ionic-covalent
Niter	$K[NO_3]$	Ionic-covalent
Cubanite	$CuFe_2S_3$	Metallic-covalent

one face requires that all faces be centered, but in the orthorhombic system, for example, centering the *C*-face does not require centering of the *A*- and *B*-faces, which are not equivalent.

A *space group* is an extended network of reflection planes, glide planes, rotation axes, inversion axes, and screw axes, all based on a system of translation of the space lattice elements. These translational elements relate identical points in space. Any symmetry operation or translation brings all the others into coincidence. Each point in the structure need not be brought back to its original position, but only to a similar position in another unit cell of the lattice.

Familiarity with space group notation and the recognition of various symmetry elements represented on drawings, such as unit cell projections of atomic positions, and on models of crystal structures, are an important part of the study of crystal chemistry. Table 6.3 lists some of the more common space groups in

Figure 6.7. Unit cell projections of atomic positions, glide and mirror planes in forsterite, Mg_2SiO_4.

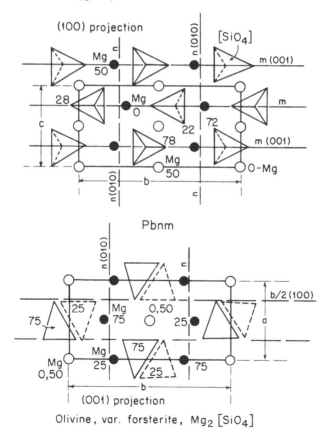

which a large number of the more common minerals crystallize. It illustrates the sequence of presentation of symmetry elements in the seven crystal systems and gives the criteria requisite for assigning a particular space group symbol or label to a particular crystal system.

Several of the 230 space groups are represented by numerous minerals and synthetic compounds, others by one or a few, and still others, though mathematically possible, are not represented by minerals at all. For example, the very common orthorhombic space group *Pbnm* is represented by many minerals embracing a wide variety of chemical composition and chemical bonding, as well (Table 6.4). In the orthorhombic crystal system, where crystal axes *a, b,*

Table 6.5. *The 230 space groups*

Point-group	Hermann–Mauguin symbols†		Schoenflies symbols
Triclinic			
1	$P1$		C_1^1
$\bar{1}$	$P\bar{1}$		C_i^1, S_2^1
Monoclinic‡	*1st setting (z-axis unique)*	*2nd setting (y-axis unique)*	
2	$P2$	$P2$	C_2^1
	$P2_1$	$P2_1$	C_2^2
	$B2.\ A2,\ I2$	$C2.\ A2,\ I2$	C_2^3
m	Pm	Pm	C_s^1
	$Pb.\ Pa,\ Pn$	$Pc.\ Pa,\ Pn$	C_s^2
	$Bm.\ Am,\ Im$	$Cm.\ Am,\ Im$	C_s^3
	$Bb.\ Aa,\ Ia$	$Cc.\ Aa,\ Ia$	C_s^4
2/m	$P2/m$	$P2/m$	C_{2h}^1
	$P2_1/m$	$P2_1/m$	C_{2h}^2
	$B2/m.\ A2/m,\ I2/m$	$C2/m.\ A2/m,\ I2/n$	C_{2h}^3
	$P2/b.\ P2/a,\ P2/n$	$P2/c.\ P2/a,\ P2/n$	C_{2h}^4
	$P2_1/b.\ P2_1/a,\ P2_1/n$	$P2_1/c.\ P2_1/a,\ P2_1/n$	C_{2h}^5
	$B2/b.\ A2/a,\ I2/a$	$C2/c.\ A2/a,\ I2/a$	C_{2h}^6
Orthorhombic	*Normal setting*	*Other settings*	
222	$P222$		$D_2^1,\ V^1$
	$P222_1$	$P2_122,\ P22_12$	$D_2^2,\ V^2$
	$P2_12_12$	$P22_12_1,\ P2_122_1$	$D_2^3,\ V^3$
	$P2_12_12_1$		$D_2^4,\ V^4$
	$C222_1$	$A2_122,\ B22_12$	$D_2^5,\ V^5$
	$C222$	$A222,\ B222$	$D_2^6,\ V^6$
	$F222$		$D_2^7,\ V^7$
	$I222$		$D_2^8,\ V^8$
	$I2_12_12_1$		$D_2^9,\ V^9$

† The Hermann–Mauguin symbols given first are those given in *International Tables* (1952).

and *c* are interchangeable and have in the past been arbitrarily selected, it is possible to transpose these axes and, in so doing, transpose the space group: there are alternatives. Space group *Pbnm* is an alternative to *Pnma* or to *Pcmn*: they are thus equivalent from an overall standpoint of symmetry. Recent crys-

Table 6.5. *(cont.)*

Point-group	Hermann–Mauguin symbols		Schoenflies symbols
	Normal	*Other settings*	
*mm*2	$Pmm2$ (Pmm)	$P2mm, Pm2m$	C_{2v}^{1}
	$Pmc2_1$ (Pmc)	$P2_1ma, Pb2_1m, Pm2_1b, Pcm2_1, P2_1am$	C_{2v}^{2}
	$Pcc2$ (Pcc)	$P2aa, Pb2b, P2aa$	C_{2v}^{3}
	$Pma2$ (Pma)	$P2mb, Pc2m, Pm2a, Pbm2, P2cm$	C_{2v}^{4}
	$Pca2_1$ (Pca)	$P2_1ab, Pc2_1b, Pb2_1a, Pbc2_1, P2_1ca$	C_{2v}^{5}
	$Pnc2$ (Pnc)	$P2na, Pb2n, Pn2b, Pcn2, P2an$	C_{2v}^{6}
	$Pmn2_1$ (Pmn)	$P2_1mn, Pn2_1m, Pm2_1n, Pnm2_1, P2_1nm$	C_{2v}^{7}
	$Pba2$ (Pba)	$P2cb, Pc2a, Pba2, P2cb$	C_{2v}^{8}
	$Pna2_1$ (Pna)	$P2_1nb, Pc2_1n, Pn2_1a, Pbn2_1, P2_1cn$	C_{2v}^{9}
	$Pnn2$ (Pnn)	$P2nn, Pn2n$	C_{2v}^{10}
	$Cmm2$ (Cmm)	$A2mm, Bm2m$	C_{2v}^{11}
	$Cmc2_1$ (Cmc)	$A2_1ma, Bb2_1m, Bm2_1b, Ccm2_1, A2_1am$	C_{2v}^{12}
	$Ccc2$ (Ccc)	$A2aa, Bb2b$	C_{2v}^{13}
	$Amm2$ (Amm)	$B2mm, Cm2m, Am2m, Bmm2, C2mm$	C_{2v}^{14}
	$Abm2$ (Abm)	$B2cm, Cm2a, Ac2m, Bma2, C2mb$	C_{2v}^{15}
	$Ama2$ (Ama)	$B2mb, Cc2m, Am2a, Bbm2, C2cm$	C_{2v}^{16}
	$Aba2$ (Aba)	$B2cb, Cc2a, Ac2a, Bb2a, C2cb$	C_{2v}^{17}
	$Fmm2$ (Fmm)	$F2mm, Fm2m$	C_{2v}^{18}
	$Fdd2$ (Fdd)	$F2dd, Fd2d$	C_{2v}^{19}
	$Imm2$ (Imm)	$I2mm, Im2m$	C_{2v}^{20}
	$Iba2$ (Iba)	$I2cb, Ic2a$	C_{2v}^{21}
	$Ima2$ (Ima)	$I2mb, Ic2m, Im2a, Ibm2, I2cm$	C_{2v}^{22}
mmm	$Pmmm$		D_{2h}^{1}, V_h^{1}
	$Pnnn$		D_{2h}^{2}, V_h^{2}
	$Pccm$	$Pmma, Pbmb$	D_{2h}^{3}, V_h^{3}
	$Pban$	$Pncb, Pcna$	D_{2h}^{4}, V_h^{4}
	$Pmma$	$Pbmm, Pmcm, Pmam, Pmmb, Pcmm$	D_{2h}^{5}, V_h^{5}
	$Pnna$	$Pbnn, Pncn, Pnan, Pnnb, Pcnn$	D_{2h}^{6}, V_h^{6}
	$Pmna$	$Pbmn, Pncm, Pman, Pnmb, Pcnm$	D_{2h}^{7}, V_h^{7}
	$Pcca$	$Pbaa, Pbcb, Pbab, Pccb$	D_{2h}^{8}, V_h^{8}
	$Pbam$	$Pmcb, Pcma$	D_{2h}^{9}, V_h^{9}
	$Pccn$	$Pnaa, Pbnb$	D_{2h}^{10}, V_h^{10}
	$Pbcm$	$Pmca, Pbma, Pcmb, Pcam, Pmab$	D_{2h}^{11}, V_h^{11}
	$Pnnm$	$Pmnn, Pnmn$	D_{2h}^{12}, V_h^{12}
	$Pmmn$	$Pnmm, Pmnm$	D_{2h}^{13}, V_h^{13}
	$Pbcn$	$Pnca, Pbna, Pcnb, Pcan, Pnab$	D_{2h}^{14}, V_h^{14}
	$Pbca$	$Pcab$	D_{2h}^{15}, V_h^{15}
	$Pnma$	$Pbnm, Pmcn, Pnam, Pmnb, Pcmn$	D_{2h}^{16}, V_h^{16}
	$Cmcm$	$Amma, Bbmm, Bmmb, Ccmm, Amam$	D_{2h}^{17}, V_h^{17}
	$Cmca$	$Abma, Bbcm, Bmab, Ccmb, Acam$	D_{2h}^{18}, V_h^{18}
	$Cmmm$	$Ammm, Bmmm$	D_{2h}^{19}, V_h^{19}
	$Cccm$	$Amaa, Bbmb$	D_{2h}^{20}, V_h^{20}
	$Cmma$	$Abmm, Bmcm, Bmam, Cmmb, Acmm$	D_{2h}^{21}, V_h^{21}
	$Ccca$	$Abaa, Bbcb, Bbab, Cccb, Acaa$	D_{2h}^{22}, V_h^{22}
	$Fmmm$		D_{2h}^{23}, V_h^{23}
	$Fddd$		D_{2h}^{24}, V_h^{24}
	$Immm$		D_{2h}^{25}, V_h^{25}

tallographic convention calls for making the longest axis b and the shortest c, but this is by no means widely accepted.

One of the many common minerals that crystallize with *Pbnm* symmetry is forsterite, the Mg end member of the olivine series. Unit cell projections of the

Table 6.5. *(cont.)*

Point-group	Hermann–Mauguin symbols		Schoenflies symbols
	Normal	*Other settings*	
mmm (cont.)	*Ibam*	*Imcb, Icma*	D_{2h}^{26}, V_{h}^{26}
	Ibca	*Icab*	D_{2h}^{27}, V_{h}^{27}
	Imma	*Ibmm, Imcm, Imam, Immb, Icmm*	D_{2h}^{28}, V_{h}^{28}
Tetragonal	*Normal*	*Larger cell*	
4	*P4*	*C4*	C_4^1
	P4$_1$	*C4$_1$*	C_4^2
	P4$_2$	*C4$_2$*	C_4^3
	P4$_3$	*C4$_3$*	C_4^4
	I4	*F4*	C_4^5
	I4$_1$	*F4$_1$*	C_4^6
$\bar{4}$	*P$\bar{4}$*	*C$\bar{4}$*	S_4^1
	I$\bar{4}$	*F$\bar{4}$*	S_4^2
4/*m*	*P4/m*	*C4/m*	C_{4h}^1
	P4$_2$/m	*C4$_2$/m*	C_{4h}^2
	P4/n	*C4/n*	C_{4h}^3
	P4$_2$/n	*C4$_2$/n*	C_{4h}^4
	I4/m	*F4/m*	C_{4h}^5
	I4$_1$/a	*F4$_1$/a*	C_{4h}^6
422	*P422*	*C422*	D_4^1
	P42$_1$2	*C422$_1$*	D_4^2
	P4$_1$22	*C4$_1$22*	D_4^3
	P4$_1$2$_1$2	*C4$_1$22$_1$*	D_4^4
	P4$_2$22	*C4$_2$22*	D_4^5
	P4$_2$2$_1$2	*C4$_2$22$_1$*	D_4^6
	P4$_3$22	*C4$_3$22*	D_4^7
	P4$_3$2$_1$2	*C4$_3$22$_1$*	D_4^8
	I422	*F422*	D_4^9
	I4$_1$22	*F4$_1$22*	D_4^{10}
4*mm*	*P4mm*	*C4mm*	C_{4v}^1
	P4bm	*C4mb*	C_{4v}^2
	P4$_2$cm (P4cm)	*C4$_2$mc (C4mc)*	C_{4v}^3
	P4$_2$nm (P4nm)	*C4$_2$mn (C4mn)*	C_{4v}^4
	P4cc	*C4cc*	C_{4v}^5
	P4nc	*C4cn*	C_{4v}^6
	P4$_2$mc (P4mc)	*C4$_2$cm (C4cm)*	C_{4v}^7
	P4$_2$bc (P4bc)	*C4$_2$cb (C4cb)*	C_{4v}^8
	I4mm	*F4mm*	C_{4v}^9
	I4cm	*F4mc*	C_{4v}^{10}
	I4$_1$md (I4md)	*F4$_1$dm (I4dm)*	C_{4v}^{11}
	I4$_1$cd (I4cd)	*F4$_1$dc (I4dc)*	C_{4v}^{12}
$\bar{4}$2*m*	*P$\bar{4}$2m*	*C$\bar{4}$2m*	D_{2d}^1, V_d^1
	P$\bar{4}$2c	*C$\bar{4}$c2*	D_{2d}^2, V_d^2
	P$\bar{4}$2$_1$m	*C$\bar{4}$m2$_1$*	D_{2d}^3, V_d^3
	P$\bar{4}$2$_1$c	*C$\bar{4}$c2$_1$*	D_{2d}^4, V_d^4

atoms parallel to (100) and to (001) planes are illustrated in Figure 6.7. They show the locations of b glides, $b/2$ (100); n glides, $a/2 + c/2$ (010); and mirror planes, m (001). In addition to learning how to locate the symmetry planes on such unit cell projection diagrams, readers should count all of the atoms in the

Table 6.5. *(cont.)*

Point-group	Hermann–Mauguin symbols		Schoenflies symbols
	Normal	*Larger cell*	
$\bar{4}2m$	$P\bar{4}m2$	$C\bar{4}2m$	$D_{2d}^5,\ V_d^5$
(cont.)	$P\bar{4}c2$	$C\bar{4}2c$	$D_{2d}^6,\ V_d^6$
	$P\bar{4}b2$	$C\bar{4}2b$	$D_{2d}^7,\ V_d^7$
	$P\bar{4}n2$	$C\bar{4}2n$	$D_{2d}^8,\ V_d^8$
	$I\bar{4}m2$	$F\bar{4}2m$	$D_{2d}^9,\ V_d^9$
	$I\bar{4}c2$	$F\bar{4}2c$	$D_{2d}^{10},\ V_d^{10}$
	$I\bar{4}2m$	$F\bar{4}m2$	$D_{2d}^{11},\ V_d^{11}$
	$I\bar{4}2d$	$F\bar{4}d2$	$D_{2d}^{12},\ V_d^{12}$
$4/mmm$	$P4/mmm$	$C4/mmm$	D_{4h}^1
	$P4/mcc$	$C4/mcc$	D_{4h}^2
	$P4/nbm$	$C4/amb$	D_{4h}^3
	$P4/nnc$	$C4/acn$	D_{4h}^4
	$P4/mbm$	$C4/mmb$	D_{4h}^5
	$P4/mnc$	$C4/mcn$	D_{4h}^6
	$P4/nmm$	$C4/amm$	D_{4h}^7
	$P4/ncc$	$C4/acc$	D_{4h}^8
	$P4_2/mmc\ (P4/mmc)$	$C4_2/mcm\ (C4mcm)$	D_{4h}^9
	$P4_2/mcm\ (P4/mcm)$	$C4_2/mmc\ (C4/mmc)$	D_{4h}^{10}
	$P4_2/nbc\ (P4/nbc)$	$C4_2/acb$	D_{4h}^{11}
	$P4_2/nnm\ (P4/nnm)$	$C4_2/amn$	D_{4h}^{12}
	$P4_2/mbc\ (P4/mbc)$	$C4_2/mcb\ (C4/mcb)$	D_{4h}^{13}
	$P4_2/mnm\ (P4/mnm)$	$C4_2/mmn\ (C4/mmn)$	D_{4h}^{14}
	$P4_2/nmc\ (P4/nmc)$	$C4_2/acm$	D_{4h}^{15}
	$P4_2/ncm\ (P4/ncm)$	$C4_2/amc$	D_{4h}^{16}
	$I4/mmm$	$F4/mmm$	D_{4h}^{17}
	$I4/mcm$	$F4/mmc$	D_{4h}^{18}
	$I4_1/amd\ (I4/amd)$	$F4_1/ddm$	D_{4h}^{19}
	$I4_1/acd\ (I4/acd)$	$F4_1/ddc$	D_{4h}^{20}
Trigonal and Hexagonal			
3	$P3\ (C3)$	$(H3)$	C_3^1
	$P3_1\ (C3_1)$	$(H3_1)$	C_3^2
	$P3_2\ (C3_2)$	$(H3_2)$	C_3^3
	$R3$		C_3^4
$\bar{3}$	$P\bar{3}\ (C\bar{3})$	$(H\bar{3})$	$C_{3i}^1,\ S_6^1$
	$R\bar{3}$		$C_{3i}^2,\ S_6^2$
32	$P312\ (C312)$	$(H32)$	D_3^1
	$P321\ (C32)$	$(H312)$	D_3^2
	$P3_112\ (C3_112)$	$(H3_12)$	D_3^3
	$P3_121\ (C3_12)$	$(H3_112)$	D_3^4
	$P3_212\ (C3_212)$	$(H3_22)$	D_3^5
	$P3_221\ (C3_22)$	$(H3_212)$	D_3^6
	$R32$		D_3^7
$3m$	$P3m1\ (C3m)$	$(H31m)$	C_{3v}^1
	$P31m\ (C31m)$	$(H3m)$	C_{3v}^2

unit cell to get the unit cell formula, to connect appropriate atom centers in order to locate octahedral and tetrahedral groups and their linkages, to derive EBS distributions, and to look at the disposition and number of shared edges between polyhedra. Ball or packing models can then be built from these atomic maps, giving a better idea of atom or ion packing. Crystal chemistry then comes alive.

A tabulation of the 230 space groups is reproduced in Table 6.5.

Table 6.5. *(cont.)*

Point-group	Hermann–Mauguin symbols		Schoenflies symbols
	Normal	*Larger cell*	
$3m$ (cont.)	$P3c1\ (C3c)$	$(H31c)$	C_{3v}^3
	$P31c\ (C31c)$	$(H3c)$	C_{3v}^4
	$R3m$		C_{3v}^5
	$R3c$		C_{3v}^6
$\bar{3}m$	$P\bar{3}1m\ (C\bar{3}1m)$	$(H\bar{3}m)$	D_{3d}^1
	$P\bar{3}1c\ (C\bar{3}1c)$	$(H\bar{3}c)$	D_{3d}^2
	$P\bar{3}m1\ (C\bar{3}m)$	$(H\bar{3}1m)$	D_{3d}^3
	$P\bar{3}c1\ (C\bar{3}c)$	$(H\bar{3}1c)$	D_{3d}^4
	$R\bar{3}m$		D_{3d}^5
	$R\bar{3}c$		D_{3d}^6
6	$P6\ (C6)$	$(H6)$	C_6^1
	$P6_1\ (C6_1)$	$(H6_1)$	C_6^2
	$P6_5\ (C6_5)$	$(H6_5)$	C_6^3
	$P6_2\ (C6_2)$	$(H6_2)$	C_6^4
	$P6_4\ (C6_4)$	$(H6_4)$	C_6^5
	$P6_3\ (C6_3)$	$(H6_3)$	C_6^6
$\bar{6}$	$P\bar{6}\ (C\bar{6})$	$(H\bar{6})$	C_{3h}^1
$6/m$	$P6/m\ (C6/m)$	$(H6/m)$	C_{6h}^1
	$P6_3/m\ (C6_3/m)$	$(H6_3/m)$	C_{6h}^2
622	$P622\ (C62)$	$(H62)$	D_6^1
	$P6_122\ (C6_12)$	$(H6_12)$	D_6^2
	$P6_522\ (C6_52)$	$(H6_52)$	D_6^3
	$P6_222\ (C6_22)$	$(H6_22)$	D_6^4
	$P6_422\ (C6_42)$	$(H6_42)$	D_6^5
	$P6_322\ (C6_32)$	$(H6_32)$	D_6^6
$6mm$	$P6mm\ (C6mm)$	$(H6mm)$	C_{6v}^1
	$P6cc\ (C6cc)$	$(H6cc)$	C_{6v}^2
	$P6_3cm\ (C6_3cm)$	$(H6_3mc)$	C_{6v}^3
	$P6_3mc\ (C6_3mc)$	$(H6_3cm)$	C_{6v}^4
$\bar{6}m2$	$P\bar{6}m2\ (C\bar{6}m)$	$(H\bar{6}2m)$	D_{3h}^1
	$P\bar{6}c2\ (C\bar{6}c)$	$(H\bar{6}2c)$	D_{3h}^2
	$P\bar{6}2m\ (C\bar{6}2m)$	$(H\bar{6}m)$	D_{3h}^3
	$P\bar{6}2c\ (C\bar{6}2c)$	$(H\bar{6}c)$	D_{3h}^4
$6/mmm$	$P6/mmm\ (C6/mmm)$	$(H6/mmm)$	D_{6h}^1
	$P6/mcc\ (C6/mcc)$	$(H6/mcc)$	D_{6h}^2
	$P6_3/mcm\ (C6/mcm)$	$(H6/mmc)$	D_{6h}^3
	$P6_3/mmc\ (C6/mmc)$	$(H6/mcm)$	D_{6h}^4

Summary

Crystals are solids enclosing three-dimensional lattices of points. These lattices are built from five two-dimensional lattices combined perpendicularly to form fourteen three-dimensional arrays of points, the Bravais space lattices.

Seven primitive lattices represent the seven crystal systems.

In minerals, the smallest coplanar array of points that translate in three-dimensional space defines the unit cell. Thus, all minerals crystallize in one each of the 7 crystal systems, 32 crystal classes, and 230 space groups.

Table 6.5. *(cont.)*

Point-group	Hermann–Mauguin symbol	Schoenflies symbol
Cubic		
23	$P23$	T^1
	$F23$	T^2
	$I23$	T^3
	$P2_13$	T^4
	$I2_13$	T^5
$m3$	$Pm3$	T_h^1
	$Pn3$	T_h^2
	$Fm3$	T_h^3
	$Fd3$	T_h^4
	$Im3$	T_h^5
	$Pa3$	T_h^6
	$Ia3$	T_h^7
432	$P432\ (P43)$	O^1
	$P4_232\ (P4_23)$	O^2
	$F432\ (F43)$	O^3
	$F4_132\ (F4_13)$	O^4
	$I432\ (I43)$	O^5
	$P4_332\ (P4_33)$	O^6
	$P4_132\ (P4_13)$	O^7
	$I4_132\ (I4_13)$	O^8
$\bar{4}3m$	$P\bar{4}3m$	T_d^1
	$F\bar{4}3m$	T_d^2
	$I\bar{4}3m$	T_d^3
	$P\bar{4}3n$	T_d^4
	$F\bar{4}3c$	T_d^5
	$I\bar{4}3d$	T_d^6
$m3m$	$Pm3m$	O_h^1
	$Pn3n$	O_h^2
	$Pm3n$	O_h^3
	$Pn3m$	O_h^4
	$Fm3m$	O_h^5
	$Fm3c$	O_h^6
	$Fd3m$	O_h^7
	$Fd3c$	O_h^8
	$Im3m$	O_h^9
	$Ia3d$	O_h^{10}

The external symmetry of a crystal allows us to determine its crystal system and class, but X-ray analysis of its internal symmetry is necessary to find its space lattice and space group. This is done by single crystal analysis, which can locate specific atoms in the structure, and by powder diffraction, which measures the distances between like planes of atoms.

Bibliography

Azaroff, L. V., and Buerger, M. J. (1958). *The powder method.* McGraw-Hill, New York.

Bloss, F. D. (1971). *Crystallography and crystal chemistry.* Holt, Rinehart & Winston, New York.

Bragg, W. L. (1933). *The crystalline state.* Vol. 1. *A general survey.* G. Bell, London.

Bragg, W. L., Claringbull, G. F., and Taylor, W. H. (1965). *Crystal structures of minerals. The crystalline state,* Vol. 4. Cornell University Press, Ithaca, N.Y.

Buerger, M. J. (1963). *Elementary crystallography.* Wiley, New York.

(1971). *Introduction to crystal geometry.* McGraw-Hill, New York.

International tables for X-ray crystallography, Vol. I (1952). Kynoch Press, Birmingham, England.

Klein, C., and Hurlbut, C. S., Jr. (1985), *Manual of mineralogy* (after J. D. Dana), 20th ed. Wiley, New York.

Klug, H. P., and Alexander, L. E. (1954). *X-ray diffraction procedures.* Wiley, New York.

Lipson, H. S. (1970). *Crystals and X-rays.* Wykeham, London.

7 Crystal field theory

Crystal field theory describes the net change in crystal energy resulting from the orientation of d orbitals of a transition metal cation inside a coordinating group of anions, also called ligands. Although considered to be largely electrostatic in origin, there is always a covalent contribution. Because it is simpler to treat the theory using an electrostatic model, we do so here.

Any free transition metal ion is classed as fivefold degenerate, because all five d orbitals are equal in energy. However, when the ion is placed in an octahedral field of six O^{2-} anions (the ligands), the degeneracy is lifted by a splitting of the crystal field into low and high energy levels. Consider an octahedral array of O^{2-} anions surrounding a transition metal cation. Oxygen anions will be positioned on the six corners of a regular octahedron on opposite ends of reference axes x, y, and z, analogous to the a, b, and c axes of crystals. Transition metal ion orbitals d_{xy}, d_{xz}, and d_{yz} (Fig. 7.1) each orient their four lobes of high electron density between octahedral axes x, y, and z and thus do not point directly at any of the six O^{2-} oxygen anions; d orbital electrons are not repelled, and all six anions can approach more closely to the cation, reducing the A—X (M–O) distance and contracting the octahedron. This contraction results in a lowering of crystal energy and an increase in stability. Each electron in orbitals d_{xy}, d_{xz}, and d_{yz} is lowered in energy by an amount $\Delta_0 = \frac{2}{5}$ (after Burns 1970a); it is two-fifths more stable than in the free or degenerate ion. These three low energy orbitals, d_{xy}, d_{xz}, and d_{yz}, constitute the t_{2g} level of crystal field theory.

Conversely, each electron located in the $d_{x^2-y^2}$ or d_{z^2} orbitals will be repelled, because their lobes of high electron density point directly at oxygen anions located on the octahedral corners. Accordingly, each such electron is raised in energy by an amount $\Delta_0 = \frac{3}{5}$ above that in a free degenerate ion. This higher energy state consisting of orbitals $d_{x^2-y^2}$ and d_{z^2} constitutes the e_g level of crystal field theory. Some find it easier to think of the oxygen anions as being repelled by the d orbital electrons, rather than the reverse, but the result is the same.

Now consider the electron occupancy of d orbitals of first-period transition metal ions (high spin configuration) from $_{21}Sc^{3+}$ to $_{30}Zn^{2+}$ (Table 7.1). Note that ions Sc^{3+} and Ti^{4+} have lost their d electrons in the ionization process, so $\Delta_0 = 0$. Ions Mn^{2+} and Fe^{3+} place one electron in each of the five $3d$ orbitals, resulting in an algebraic cancellation of increased and decreased energies, to a net $\Delta_0 = 0$, or 0/5; Zn^{2+} has two electrons paired in each of the five $3d$ orbitals, also effecting a cancellation of CFSE (crystal field stabilization energy) to $\Delta_0 = 0$. In all other cases, the transition metal cations of Table 7.1 will split the octahedral crystal field into low (t_{2g}) and high (e_g) energy levels and gain CFSE; each will have an octahedral site preference energy commensurate with its value of Δ_0.

It is important to see that Cr^{3+}, with three electrons in t_{2g}, will permit all six coordinating octahedral O^{2-} anions to approach closely to form a small or contracted, but regular, octahedron. In contrast, the addition of a fourth electron into d_{x2-y2} or d_{z2}, as in Mn^{3+} or Cr^{2+}, will induce negative charge repulsion along either the x-y octahedral axes or along the z axis, depending on which orbital holds the fourth d electron. Accordingly, the repulsion of negative charge will cause the octahedron to become distorted (Fig. 7.2); d_{x2-y2} occu-

Figure 7.1. When a transition metal cation lies within an octahedral crystal field formed of six O^{2-} anions or ligands, cation orbitals, d_{x2-y2} and d_{z2} orient along octahedral axes, x, y, and z, in high energy level, eg; they are repelled by the six O^{2-} ligands. Cation orbitals, d_{xy}, d_{yz}, and d_{xz} orient between octahedral axes, lie in low energy level, t_{2g}, and are attracted by, drawing closer to, the cation.

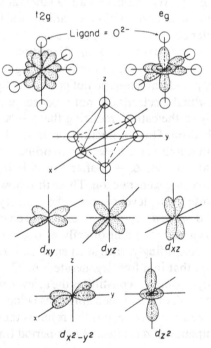

pancy will induce a tetragonal compressed or oblate octahedral distortion, whereas d_{z^2} occupancy will introduce a tetragonal elongate or prolate octahedral distortion. These distortions are classed as Jahn–Teller distortions in crystal field theory. Note that the addition of a fifth d electron, as in Mn^{2+} or Fe^{3+}, will cancel both the CFSE and the octahedral distortion.

Because crystal field theory has been successful in explaining cation ordering in the spinel mineral group, it will prove instructive to examine several examples. All members of this group have sixteen octahedral cations and eight tetrahedral cations along with thirty-two oxygen anions in one unit cell containing eight formula units ($Z = 8$). Spinel formulae confuse students because some mineralogists write $Mg^{IV}Al_2^{VI}O_4$ and others write $Al_2^{VI}[Mg^{IV}O_4]$: the latter is chosen here because it provides more crystal chemical information. The entry $[MgO_4]$ clearly informs us that the Mg cation is four-coordinated by O^{2-}, a fact often unknown in this chemically very complex spinel group.

Although many natural and synthetic compounds crystallize with the spinel structure, the most appropriate example to consider first is chromite, $Cr_2^{VI}[Fe^{IV}O_4]$. From Table 7.1 we see that Cr^{3+}, with three unpaired d electrons,

Table 7.1. *d orbital occupancy and octahedral crystal field stabilization energies for first series transition metal ions in high spin configuration*

Z	Ion	d els.	t2g (d_{xy}, d_{xz}, d_{yz})			eg ($d_{x^2y^2}$, d_{z^2})		C.F.S.E. Δ_0
21	Sc³⁺	0	○	○	○	○	○	0/5
22	Ti⁴⁺	0	○	○	○	○	○	0/5
22	Ti³⁺	1	↑	○	○	○	○	2/5
23	V³⁺	2	↑	↑	○	○	○	4/5
24	Cr³⁺	3	↑	↑	↑	○	○	6/5
24	Cr²⁺	4	↑	↑	↑	↑	○	3/5
25	Mn³⁺	4	↑	↑	↑	○	↑	3/5
25	Mn²⁺	5	↑	↑	↑	↑	↑	0/5
26	Fe³⁺	5	↑	↑	↑	↑	↑	0/5
26	Fe²⁺	6	↑↓	↑	↑	↑	↑	2/5
27	Co²⁺	7	↑↓	↑↓	↑	↑	↑	4/5
28	Ni²⁺	8	↑↓	↑↓	↑↓	↑	↑	6/5
29	Cu²⁺	9	↑↓	↑↓	↑↓	↑↓	↑	3/5
30	Zn²⁺	10	↑↓	↑↓	↑↓	↑↓	↑↓	0/5

has CFSE $\Delta_0 = \frac{8}{5}$, whereas Fe^{2+} has only four unpaired d electrons and $\Delta_0 = \frac{2}{5}$: thus Cr^{3+} enters both octahedral sites, causing the ion with the larger ionic radius, Fe^{2+}, to enter the tetrahedral site. Crystal field theory states that Cr^{3+}, $\Delta_0 = \frac{8}{5}$, gains more stabilization energy than any other competing cation through occupancy of the octahedral, rather than the tetrahedral, site.

The formula for magnetite is often written Fe_3O_4 in order to avoid more specific location of the Fe^{2+} and Fe^{3+} ions in the structure. The data of Table 7.1 show that Fe^{2+}, $\Delta_0 = \frac{2}{5}$, gains octahedral site preference over Fe^{3+}, $\Delta_0 = 0$, but complications arise because $Fe^{2+}:Fe^{3+} = 1:2$, and CN VI sites:CN IV sites $= 2:1$. Crystal field relations place the one Fe^{2+} ion in the octahedron, leaving the two Fe^{3+} ions to occupy a tetrahedral site and an octahedral site. The formula is now correctly written as $Fe^{2+VI}Fe^{3+VI}[Fe^{3+IV}O_4^{IV}]$. Magnetite is classed as an inverse spinel compared to chromite, which is a normal spinel.

Another example of the application of crystal field theory to spinels is provided by hausmannite, Mn_3O_4, which is distorted from cubic (isometric) to

Figure 7.2. Comparison of regular, prolate-distorted, and oblate-distorted octahedra. Distortions are caused by unbalanced distribution of electrons in high and low energy d orbitals of a transition metal ion.

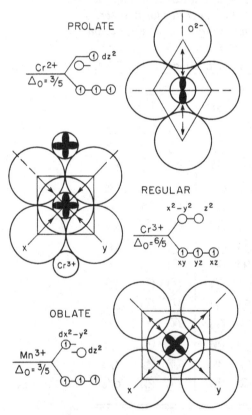

tetragonal symmetry by Jahn–Teller distortions of octahedral sites occupied by the Mn^{3+} ions. Note that Mn^{3+} has $\Delta_0 = \frac{2}{3}$ and Mn^{2+} has $\Delta_0 = 0$; thus the smaller radius Mn^{3+} is placed in the octahedral sites, and the much larger Mn^{2+} ion is placed in the tetrahedral site. Because the length of the c axis is greater than that of the a axis, the distortion is assumed to be prolate, with the fourth electron in Mn^{3+} entering d_{z^2}.

The mineral spinel for which the group is named (see Chapter 18) has the formula $Al_2^{VI}[Mg^{IV}O_4^{IV}]$, and neither Al^{3+} nor Mg^{2+} possesses any d orbital electrons or any crystal field stabilization energy. Why, then, does the smaller radius Al^{3+} ion occupy both octahedral sites, and the larger Mg^{2+} ion the tetrahedral site? It is a general truth of crystal chemistry that reduced inter-atomic distances, especially for ions with high formal charge, lead to increased stability; so Al^{3+} is favored in the octahedral sites and is said to have higher lattice energy. Such lattice energy must be very comparable to crystal field energy in magnitude because the synthetic Al_2NiO_4 spinel is said by McClure (1957) to be very disordered, with roughly equal amounts of Al^{3+} and Ni^{2+}, $\Delta_0 = \frac{4}{5}$, in both octahedral and tetrahedral sites. All other Al-rich spinels show the Al^{3+} ion occupying both octahedral sites. Burns (1970b) has estimated that $\Delta_0 = 1/(M\text{–}O \text{ distance})^5$, or that CFSE is inversely proportional to interatomic distance, so the smaller octahedral site will always have a greatly increased CFSE.

Location of a transition metal inside a coordination polyhedron other than an octahedron can also confer CFSE on a crystal, but d orbital lobes that point to corners of an octahedron do not point toward the corners of either a cube or a tetrahedron. Again, however, energy is lowered in these CN VIII and CN IV polyhedra when d orbital lobes point away from anions, and raised when they point at or near to the anions disposed on corners of these polyhedra. The disposition of these lobes of electron density with respect to the geometric location of oxygen anions confers lower CFSE values for cubic coordination than for octahedral coordination, and still lower values for tetrahedral coordination. Because of differing geometries, the low t_{2g} and high e_g energy levels of the octahedron are reversed in both the cube and the tetrahedron. Try to draw a cube or a tetrahedron around the lobes of the five d orbitals, and it will be apparent that CFSE energies will be lower because of the difficulty in orienting the lobes as efficiently as can be done in the octahedron.

According to Hund's classical selection rules (see Chapter 1), transition metal electrons should occupy d orbitals singly before doubly, favoring the high spin over the low spin state. This appears to be the rule in most oxide and silicate minerals. In some cases, however, spin pairing may lead to a lower crystal energy than would a high spin configuration. In many sulfides, Fe^{2+} and Co^{2+} ions adopt the low spin configuration. Spin pairing of d orbital electrons leads to increased π bond formation (see Chapter 3), and CFSE, which in turn reduces interatomic distances (M–S) and strengthens metal–sulfur bonds. This once again reinforces our suggestion that a covalent component is always to be expected in so-called ionic compounds. Where π bonds form between transition metal and sulfur, a molecular orbital extends throughout the crystal, endowing it with metallic luster and electrical conductivity.

Summary

Crystal field theory describes the net change in crystal energy resulting from the orientation of d orbitals of a transition metal cation inside a coordinating group of anions (ligands).

Inside an octahedron of oxygen anions, a high spin transition metal ion (fivefold degenerate in the free state, because its five d orbitals are equal in energy) has its valence electrons split into three low energy orbitals, d_{xy}, d_{xz}, d_{yz}, and two high energy orbitals, $d_{x^2-y^2}$, d_{z^2}. Ions Sc^{3+} and Ti^{4+} lose their d electrons in ionizing and have $\Delta_0 = 0$. Mn^{2+} and Fe^{2+}, with one electron in each of five $3d$ orbitals, cancel high and low energy levels so that $\Delta_0 = 0$, as does Zn^{2+} with two electrons paired in each of the five d orbitals. All other transition metal cations gain crystal field stabilization energy, and those with d^4 or d^9 configurations (e.g., Mn^{3+} and Cu^{2+}) cause Jahn–Teller distortions of the octahedron.

Crystal field relations cause various distortions in minerals crystallizing with the spinel structure and may be a dominant factor in cation ordering. In polyhedra other than octahedra, crystal field stabilization energy of d orbitals is lowered when the orbital lobes point away from anions and raised when they point toward anions.

CFSE values are lower in magnitude in cubic and tetrahedral coordination than in octahedral cation coordination.

In sulfide minerals, spin pairing of d orbital electrons leads to lower crystal energy through increase in both π bond formation and CFSE (Fe^{2+} in low-spin configuration will have $\Delta_0 = \frac{12}{5}$ because all six d electrons are paired in the three low energy orbitals).

Bibliography

Ballhausen, C. J. (1962). *Introduction to ligand field theory.* McGraw-Hill, New York.

Burns, R. G. (1970a). *Mineralogical applications of crystal field theory.* Cambridge Univ. Press, London.

(1970b). Crystal field spectra and evidence of cation ordering in olivine minerals. *Am. Mineral.,* 55:1608–32.

Cotton, F. A., and Wilkinson, G. (1966). *Advanced inorganic chemistry,* 2d ed. Interscience, New York.

Dunitz, J.D., and Orgel, L. E. (1957). Electronic properties of transition metal oxides. II. Cation distribution amongst octahedral and tetrahedral sites. *J. Phys. Chem. Solids,* 3:318–33.

McClure, D. S. (1957). The distribution of transition metal cations in spinels. *J. Phys. Chem. Solids,* 3:311–17.

8   Polyhedral distortion

The tetrahedron, octahedron, and cube are geometric solids that possess the basic external symmetry elements required by the isometric crystal system, specifically four threefold axes of rotational symmetry, whereas planar triangles and squares will possess one threefold and one fourfold rotation axis, respectively. Close examination of the geometry of these polyhedral or planar forms by single crystal X-ray analyses of crystals shows that these requirements are rarely satisfied. Atomic groupings in most minerals are considerably distorted, and, strictly speaking, isometric tetrahedra, octahedra, and cubes rarely occur. Regular, symmetrically disposed groupings of atoms, however, are to be expected in those crystals whose atoms are polymerized by the use of covalent hybrid bond orbitals such as sp^2 (120°) planar, equilateral triangular as in graphite; sp^3 (109°28′) regular tetrahedral as in diamond; dsp^2 (90°) regular, square planar (or rectangular) as in cooperite (PtS); d^2sp^3 (90°) regular octahedral orbitals of Fe combined with S—S sp^3 orbitals as in pyrite.

Those crystals in which ionic bonding predominates will have their symmetry distorted by local imbalance of electrical charge, by cation repulsion across a shared polyhedral edge or bridge, or by Jahn–Teller distortions of crystal field theory. In the pyroxene diopside (see Fig. 5.6, Table 5.1), it was noted that unequal EBS distribution over three different oxygens was associated with unequal distances in Si—O_4 tetrahedra, Mg—O_6 octahedra, and in Ca—O_8 square antiprisms; the oxygen polyhedra are distorted as shown (Fig. 8.1), and the cations are not centrosymmetrically oriented in them. Such phenomena are also found in the magnesian olivine forsterite (Fig. 8.2).

In the idealized structure of olivine, the oxygen anions form a perfect hexagonal close-packed array, with Mg ions in regular octahedra and Si ions in regular tetrahedra. The actual structure is not at all close packed but is built of kinked Mg—O octahedral chains that run parallel to the c axis in a zigzag pattern. The chains are built of two types of polyhedra: elongated Mg—O_6 octahedra designated M-1 that run zigzag parallel to c, and alternately attached extra links representing a larger octahedron designated M-2 (Fig. 8.2). Each oxygen in the structure links two M-1 octahedra to one M-2 octahedron and one tetrahedron. Both the drawings and the interatomic distances for O—O, Si—O, and Mg—O reveal large distortions from isometric coordination polyhedra. Here, the cause of these distortions lies in the repulsion of positive charges on Mg^{2+} and Si^{4+} (ideal charges) cations that approach too closely on opposite sides of O—O bridges. Note that each tetrahedron shares three of its six edges with octahedra, each M-1 octahedron shares six of its twelve edges with M-2 octahedra and tetrahedra, and each M-2 octahedron shares three of its six edges with the other polyhedra.

Shared edges, O—O, shorten from an ideal 2.80-Å distance to 2.56 Å, and unshared edges lengthen to 3.39 Å, resulting in the considerable distortion of all polyhedra. Regular and distorted octahedra are compared in packing models (Fig. 8.3), where $Mg—O_6$ octahedra in periclase are regular because all twelve edges are shared, eliminating the possibility of electrostatically induced distortion; $Al—O_6$ "octahedra" in corundum are distorted to trigonal antiprisms because Al—Al repulsion occurs across one shared face and not the others (also see Figs. 18.2, 18.3); and M-1 and M-2 octahedra in forsterite show the distortions described above. Thus, in periclase, the oxygen anions are in cubic close packing, where all octahedral edges are shared; whereas, in both forsterite and corundum the oxygens only approximate a hexagonal close-packed array. Cur-

Figure 8.1. Comparison of regular with distorted square antiprismatic, octahedral, and tetrahedral polyhedra in diopside.

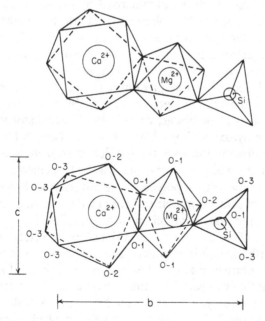

iously, the electrostatic phenomenon of cation–cation repulsion that induces the shared-edge shortening results in O—O overlap, a covalent phenomenon.

Electrostatically induced distortion of polyhedra also occurs where crystal field splitting of octahedral sites places a fourth or ninth d electron in either one of the axially directed orbitals $d_{x^2-y^2}$ or d_{z^2} (see Chapter 7) and causes the prolate octahedral distortion observed in hausmannite, $Mn_2^{3+}[Mn_2O_4]$, which then assumes a tetragonal symmetry rather than the isometric symmetry of a normal spinel.

Many minerals in which Cu^{2+}, $3d^9$, is bonded to oxygen show characteristic square planar arrays of O around Cu, with fifth and sixth oxygens at corners of very distorted, prolate or elongated, "octahedra." Here, as elsewhere in crystal chemistry, two interpretations are possible. The crystal field argument would be based on the unbalanced electron distribution that places paired electrons in $d_{x^2-y^2}$ and a single electron in d_{z^2}; an alternative would be the formation of a dsp^2 square planar hybrid (see Chapter 3) on Cu^{2+} in which the four orbitals of the hybrid point toward corners of a square; the ninth electron, not clearly accounted for, may be involved in π bonding.

The same arguments may be offered for the compound CrF_2, where Cr^{2+} has a $3d^4$ configuration; the prolate octahedral distortion observed could result from placement of the fourth d electron in d_{z^2} (Jahn–Teller distortion) or by the formation of a square planar hybrid bond orbital on Cr^{2+}. In each example, the problem might be solved by determining paramagnetic susceptibilities: high magnetic response would indicate unpaired electrons and favor the crystal field

Figure 8.2. Distortion of octahedra and tetrahedra in forsterite, Mg_2SiO_4, induced by shared-edge shortening.

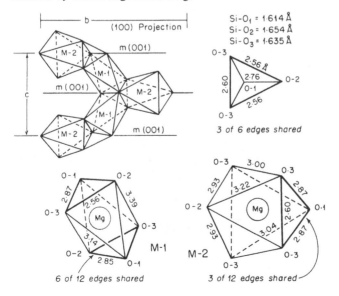

98   *Principles of crystal chemistry and refractivity*

argument; low magnetic response would indicate electron pairing and favor the hybrid bond orbital argument. Pyrite, FeS_2, and hauerite, MnS_2, though they are isostructural, show totally different magnetic properties. Pyrite is essentially diamagnetic, and hauerite is strongly paramagnetic; d electrons of Fe^{2+} in pyrite must be paired in low spin configuration, whereas d electrons of Mn^{2+} in hauerite must be unpaired or in high spin configuration (see Chapter 3).

Some minerals have their oxygen atoms linked to other atoms by both covalent and ionic bonds. In barite, $Ba[SO_4]$ (Colville and Staudhammer 1967), sp^3 hybrid covalent bonds link S to O in $[SO_4]^{2-}$ groups, and these groups in turn are held together by Ba—O bonds of predominantly ionic character. In barite, the large-radius Ba^{2+} ion ($r_{Ba} = 1.60$ Å) is coordinated by twelve oxygen atoms or anions that form parts of seven different $[SO_4]^{2-}$ tetrahedra. Cation repulsion across the large number of O—O edges shared between covalent $[SO_4]$ and ionic $Ba—O_{12}$ polyhedra can perturb and distort the otherwise more regular $[SO_4]$ groups. Pauling (1960) has noted that the $[SO_4]$ group, although built predominantly of sp^3 hybrid covalent bonds, also has a small ionic component, and the Ba—O bonds must, in turn, have a covalent component of 17%, calculated from electronegativity differences. Repeated here, for emphasis, is Pauling's warning that, although most atoms are charged, they carry a far lower charge than is indicated by formal oxidation states.

To sum up, it appears that polyhedral distortion in crystals results from a variety of crystal chemical phenomena that are essentially electrostatic in ori-

Figure 8.3. Packing models comparing a regular octahedron (A) as found in periclase, with distorted octahedra (B) in corundum, (C) in the M-1 site of olivine, and (D) in the M-2 site of olivine.

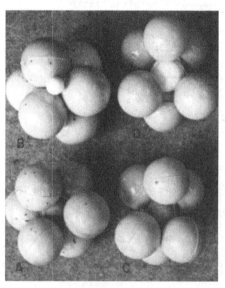

gin. These phenomena include ionization, cation repulsion across shared anion bridges such as O—O, local imbalance of electrostatic bond strength (uneven charge distribution), and Jahn–Teller distortions associated with unbalanced occupancy of axially directed d_{z^2} and $d_{x^2-y^2}$ orbitals in octahedral fields.

Polyhedral regularity of geometry is to be expected in crystals polymerized essentially by directed hybrid bond covalent orbitals, sp^2, sp^3, dsp^2, and d^2sp^3.

Summary

Strictly isometric coordination polyhedra rarely occur in crystals in which ionic bonding predominates, though they may be found in crystals polymerized by covalent hybrid bond orbitals. Crystals with predominantly ionic bonding, local electrostatic imbalance, cation repulsion across a shared edge or bridge, or Jahn–Teller effects contain distorted polyhedra. Examples of each of these cases are discussed.

Bibliography

Birle, J. D., Gibbs, G. V., Moore, P. B., and Smith, J. V. (1968). Crystal structures of natural olivines. *Am. Mineral.,* 53:807–24.

Belov, N. V., Belova, E. N., Andrianova, N. N., and Smirova, R. F. (1951). Parameters of olivine. *Dokl. Akad. Nauk SSSR,* 81:399.

Bragg, L., Claringbull, G. F., and Taylor, W. H. (1965). *Crystal structures of minerals. The crystalline state,* Vol. IV. Cornell Univ. Press, Ithaca, N.Y.

Burns, R. G. (1970). *Mineralogical applications of crystal field theory.* Cambridge University Press, London.

Brown, G. E., Jr. (1982). Olivines and silicate spinels. *Mineral. Soc. Am. Rev. Mineral.,* 5:275–381.

Colville, A. A., and Staudhammer, K. (1967). A refinement of the structure of barite. *Am. Mineral.,* 52:1877–80.

Hazen, R. M., and Finger, L. W. (1985). Crystals at high pressure. *Sci. Am.,* 252:110–17.

Pauling, L. (1960). *The nature of the chemical bond,* 3d ed. Cornell Univ. Press, Ithaca, N.Y.

9   Diadochy and isostructural crystals

The proliferation of terms used to describe aspects of substitution of ions or atoms in crystals includes diadochy, isomorphism, isostructuralism, isotypism, limited substitution, and solid solution. Because these terms have been inconsistently used, they are defined here in terms of their etymology.

Diadochy (from Greek, $\delta\iota\alpha\delta\epsilon\chi o\mu\alpha\iota$, to receive from one another, to succeed, to substitute for) is obviously the most appropriate term to use for the substitution of atoms in crystals. As used here, it covers the limited substitution of ions in a particular mineral species, as well as complete solid solution of two or more ions in related mineral species.

Isomorphism (from Greek, $\iota\sigma o\mu o\rho\varphi o\varsigma$, equal in form) is intended to relate a group of minerals of similar or identical morphology and analogous formula that contain different ions in a given coordination site; for example, calcite, $Ca^{VI}[CO_3]$, magnesite, $Mg^{VI}[CO_3]$, rhodochrosite, $Mn^{VI}[CO_3]$, and siderite, $Fe^{VI}[CO_3]$. The terms *isomorphous* or *isomorphism* have been incorrectly used to imply that the group shows solid solution relations. Ca and Mg of calcite and magnesite, respectively, for example, show only very limited solid solution relations under natural conditions, whereas the others show extensive to complete solid solution.

Isostructural (Greek $\iota\sigma o$ + Latin *structura*, meaning equal in structure). As stringently defined here, isostructural crystals must exhibit equivalence of coordination numbers of all cations and anions, space group, formula, and the number of formulae *(Z)* or atoms in one unit cell. It clearly does not imply solid solution relations, and frequently none exist.

The terms *limited solid solution* and *complete solid solution* are treated here as aspects of diadochy, and isomorphism is abandoned in favor of isostructuralism, although only nuances separate them.

Diadochy and isostructuralism are subdivided in Figure 9.1, and a discussion of each type follows.

Ion–ion and coupled substitutions

The great pioneer ("father") of geochemistry and crystal chemistry was V. M. Goldschmidt. On the basis of the very limited experimental data available in 1920–40, he foresaw and formulated many principles of crystal chemistry that are still valid today. He taught that ionic radius and electric charge are principal factors controlling diadochy in minerals crystallizing from magmas. Three of his substitution categories included *camouflage, capture,* and *admission.*

Camouflage represents the substitution of one element, often a trace element, for a major element in a host mineral when both elements have similar radii ($\pm$ 15%) and identical electric charge; for example, Hf^{4+VIII} ($r = 0.83$ Å) replaces Zr^{4+VIII} ($r = 0.84$ Å) in the distorted square antiprismatic coordination site of zircon, $Zr[SiO_4]$ (Figs. 12.5, 12.6), both elements tending to concentrate in very late stages of magmatic crystallization.

Capture refers to the substitution of one minor element having similar ionic radius to, but higher electric charge than, the host element it replaces. The excess charge must be balanced by another substitution that reduces charge, and a coupled substitution often results; for example, $Th^{4+}Si^{4+} \leftrightharpoons Ce^{3+}P^{5+}$ in monazite, (Ce,La) $[PO_4]$; $Nb^{5+}Fe^{3+} \leftrightharpoons 2Ti^{4+}$ in sphene (titanite), CaTi (O,F) $[SiO_4]$; or $Ba^{2+}Al^{3+} \leftrightharpoons K^+Si^{4+}$ in orthoclase, $K[AlSi_3O_8]$.

Admission refers to the substitution of an element of similar ionic radius, but lower charge, for a host element; in the three examples used, the admission process is coupled to the capture process to balance electric charge.

Figure 9.1. Classification scheme for diadochic and isostructural compounds.

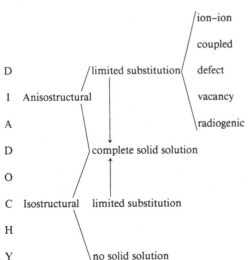

Defect or omission substitution

Defect or omission substitution may occur when a crystal structure possesses void sites that may only be partially occupied by other ions. The classic example is pyrrhotite, $Fe_{1-x}S$ or $Fe_{11}S_{12}$, in a lattice that does not have an equal amount of Fe and S atoms. This mineral is thus Fe deficient, and the lattice is said to be defective. Note that the Fe and S are covalently bonded, and electric charge need not be balanced in the same manner as in ionic structures. The common mineral magnetite, $Fe^{3+}Fe^{2+}[Fe^{3+}O_4]$, may alter by oxidation processes to maghemite (γ-Fe_2O_3), which is really an ion-deficient and iron-deficient spinel structure in which the combined three octahedral and tetrahedral sites are only partially occupied, all by trivalent Fe; the formula may be written as $[Fe_{1.78}\square_{0.22}]$ $[Fe_{0.88}\square_{0.12}O_4]$.

Another fine example of defect substitution occurs in the complex oxide mineral pyrochlore, $Na,Ca^{VIII}Nb_2^{VI}O_6^{IV}F^{IV}$, which was once considered rare but is now the principal ore mineral of niobium. When alloyed with iron, niobium produces high-quality steel. The chemical formula cited above is ideal and is seldom met with in nature. Because pyrochlore is built of an octahedral framework in which each NbO_6 octahedron shares all of its six corner oxygens with other octahedra, as much as 75% of the Na, Ca, and F ions can be removed, leaving gaping holes in a defect lattice of formula $\square_{1.5}^{VIII}(Ca,Ba)_{0.5}Nb_2^{VI}O_4^{IV}(OH)_3^{IV}$. Neutrality is maintained by a compensating substitution of $(OH)^-$ for O^{2-}, according to the scheme $NaCaO_3 \leftrightharpoons \square\square(OH)_3$.

In other varieties of pyrochlore, where the mineral is concentrated in a chemical weathering process under tropical humid conditions, Na and Ca are removed and Ba is partially substituted, leaving voids in the structure. Here we are obviously dealing with weak one-electron ionic bonds Na—O, Na—F, Ca—O, Ca—F that can be broken much more readily than the stronger, five-electron bonds Nb^{5+}—O that are undoubtedly in part covalent (see Figs. 18.18, 18.19).

Vacancy substitution

One of the most interesting types of diadochy results from the substitution or entry of large stray ions or even molecules into structural voids that exist in virtually perfect crystals as tunnels. For example, in beryl, $Al_2Be_3[Si_6O_{18}]$, the columns of this hexagonal lattice normally have empty tunnels that pass through $[Si_6O_{18}]$ groups that lie parallel to the c axis of the crystal. These tunnels are large enough in diameter to accept stray ions of Cs^+ ($r = 1.88$ Å), the largest of all cations, as well as water molecules ($r = 1.38$ Å). Still other minerals carry residual surface charges on cleavage or other faces that may attract oppositely charged ions. Another fine example of vacancy substitution occurs in amphiboles that contain a distorted tenfold coordination site that is commonly empty, for example, tremolite, $\square Ca_2Mg_5[Si_8O_{22}](OH)_2$. Here, a coupled substitution also involving these vacant sites takes place according to the scheme, $Na^{+X}Al^{3+IV} \rightleftharpoons \square^X Si^{4+IV}$.

Crystal field substitution

It should be evident that crystal field stabilization energy (see Chapter 7) may play a significant role in diadochy. A study made by Curtis (1964) showed that effective concentration of Cr^{3+} ($\Delta_0 = \frac{3}{5}$) and Ni^{2+}($\Delta_0 = \frac{2}{5}$) in place of the Mg^{2+} in octahedral sites of early formed olivine and pyroxene of the Skaergaard Complex, Greenland, could readily by explained by CFSE processes. Conversely, rejection of Mn^{3+} and Cu^{2+} from these same octahedral sites was related to their Jahn–Teller distortions and probable destabilization of octahedral sites, resulting in residual concentration of these elements in late magmatic liquids.

Radiogenic substitution

Radiogenic substitution is a complex process that may introduce a large ion into a small site because of radioactive decay and, in so doing, may result in the ultimate destruction of the ordered crystal lattice of the host mineral. For example, zircon, alluded to earlier, always carries small amounts of uranium and thorium, both of which decay, largely by α-decay processes, to lead and helium. Entry of U^{4+} ($r = 1.00$ Å) and Th^{4+} ($r = 1.04$ Å) into the CN VIII sites of Zr ($r = 0.84$ Å) results in a small lattice expansion, depending on the amount of each that is introduced. Radioactive decay of the U^{4+} and Th^{4+} to Pb^{2+} ($r = 1.32$ Å) expands the CN VIII sites and the lattice considerably. Depending on radioactive dosage (concentration + time), the zircon lattice may become disordered (Holland and Gottfried 1955), bit by bit, in a process called *metamictization,* in which water enters into radiation-induced fractures and eventually alters and destroys the lattice: the zircon becomes amorphous. Such changes in the degree of crystallinity are accompanied by a lowering of the indices of refraction that can be correlated with an increase of radiogenic lead in zircon (Gottfried, Jaffe, and Senftle 1959).

Isostructural compounds and applied crystal chemistry

To this point, we have considered diadochy largely from an ionic viewpoint, although we know from electronegativity differences that most chemical bonds in minerals carry a considerable covalent component. A covalent component, however, does not negate the ionic component. According to Pauling (1960), it merely transfers and reduces formal charges on cations so that Si^{4+} may really be Si^{1+}, but ions remain electrically charged. This reduction and transfer of charge to a resultant charge near unity embodies Pauling's electronegativity principle. How important, then, are electronegativity, ionization potential, and the chemical bond type in diadochy? Some of the answer must come from the large number of minerals and synthetic compounds that are isostructural but do not show limited substitution, nor do they form any solid solution series.

A list of $A^{VI}X^{VI}$ structures, all isostructural with halite, is given in Table 9.1, along with a comparison of their electronegativity values, percentages of ionic bonding calculated from these, and some physical and chemical properties. Eight of these occur as natural minerals, and nine are synthetic compounds. These data reveal a wide spectrum of chemical bond types, with index of refraction n and luster especially revealing. There is little evidence that any of these form solid solution series, although all have the identical crystal structure of $Na^{VI}Cl^{VI}$. Why, then, do such chemically heterogeneous compounds grow with the identical crystal structure? The important principle here is that atoms in space are disordered and can lower their energy by ordering themselves into the lattice of a growing crystal. Atoms must come into tangential ionic contact or be overlapped in covalent or metallic union, and the VI–VI coordination space lattice becomes one of cubic close packing of the larger atom, with the smaller partner filling the octahedral voids. Thus twelve large atoms surround one another, while six of these atoms surround the smaller atom, and vice versa. Our model predicts that atom pairs in these crystals fill space very economically, but it does not tell us whether the bonds are ionic, covalent, metallic, or a combination of them. We can deduce the bond type from physical properties or from electronegativity predictions. Close-packed atomic arrays form low free energy, high stability assemblages, whether the principal bond be ionic, cova-

Table 9.1. *Compounds isostructural with halite $4/Na^{VI}Cl^{VI}Fm3m$*

Compound (all VI–VI)	Name	n	% Ionic bonding	Luster	Solubility
NaF	Villiaumite	1.328	91	Vitreous	H_2O
LiF	Synthetic	1.392	70	Vitreous	H_2O
CsF	Synthetic	1.478	93	Vitreous	H_2O
KCl	Sylvite	1.490	70	Vitreous	H_2O
NaCl	Halite	1.544	67	Vitreous	H_2O
RbBr	Synthetic	1.553	63	?	H_2O
RbI	Synthetic	1.647	51	?	H_2O
LiBr	Synthetic	1.784	55	?	H_2O
LiI	Synthetic	1.955	43	?	H_2O
AgCl	Cerargyrite	2.06	26	Adamantine	$(NH_4)(OH)$
MnO	Manganosite	2.16	63	Adamantine	HCl
NiO	Bunsenite	2.23	51	Adamantine	HCl
MgS	Synthetic	2.27	34	Adamantine	PCl_3
MnS	Alabandite	Opaque	22	Submetallic	HCl
PbS	Galena	Opaque	12	Metallic	HNO_3
TiC	Synthetic	Opaque	22	Metallic	Aqua regia

lent, or metallic. Closeness of packing, irrespective of bond type, is the hallmark of stability.

Solid-solution series

Solid solution series are too well known to mineralogists and chemists to need much elaboration here. The olivine series $Mg_2SiO_4 \leftrightarrows Fe_2SiO_4$, the orthopyroxene series $Mg_2Si_2O_6 \leftrightarrows Fe_2^{VI}Si_2^{IV}O_6$ complete at high pressures, the plagioclase series $NaAlSi_3O_8 \leftrightarrows CaAl_2Si_2O_8$ at high temperature, the garnet series

$$(Mg \leftrightarrows Fe \leftrightarrows Mn \leftrightarrows Ca)_3(Al \leftrightarrows Fe^{3+} \leftrightarrows Cr^{3+})_2[SiO_4]_3$$

and many others could be cited. Here we deal with complete solid solution series and isostructural compounds as well, although some of these require high pressure or high temperature to stabilize particular members of a series.

That element concentration in magmas can restrict solid solution series that might otherwise form is exemplified by the yttrium-bearing manganese garnet spessartine. Jaffe (1951) suggested that yttrium could enter the garnet lattice by use of the coupled substitution $Y^{3+VIII}Al^{3+IV} \leftrightarrows Mn^{2+VIII}Si^{4+IV}$, and proposed the formula $(Mn_{3-x}Y_x)Al_2[Si_{3-x}Al_xO_4]_3$. He found a maximum of about 2.50% Y_2O_3 in garnet, in which $x = 0.1$ in a formula ratio of $(Mn_{2.9}Y_{0.1})$. Yoder and Keith (1951) later were able to synthesize the end member $Y_3Al_2[AlO_4]_3$, using Jaffe's substitution scheme and 1 g of Y_2O_3 purchased for ten dollars by Jaffe for the study. This modest investment paid dividends and continues to do so, since Yoder and Keith synthesized this yttrium garnet that has become known as YAG.

Introduction of Fe^{3+} in place of Al^{3+} then followed, with the growth, in the laboratory, of YIG, yttrium iron garnet. These studies, made some thirty-five years ago, showed clearly that the same element could occupy two different coordination sites in the same mineral. Al^{3+} occupies octahedral and tetrahedral sites in YAG, in sillimanite, $AlOSi[AlO_4]$, and in muscovite, $KAl_2[Si_3AlO_{10}](OH)_2$. In andalusite, Al^{3+} occupies a distorted fivefold coordination site in addition to the tetrahedral site, giving the formula $AlOSi[AlO_4]$. Synthesis of YIG provided a rare example of occupancy of both octahedral and tetrahedral sites by Fe^{3+} in the same mineral (also see Chapter 12). This also occurs in the defect lattice of maghemite described in this chapter.

Synthetic rare earth garnets, such as the prototypes YAG and YIG, have had a wide variety of uses over the years. YIG was used successfully in radar devices, and YAG has been used as a gemstone and most recently as the crystal source of the laser beam in apparatus used in eye surgery. These examples, as well as numerous synthetic gems (star sapphire, alexandrite, and diamond), synthetic zeolites used for molecular sieves in petroleum refining processes, and even a synthetic rock developed to store radioactive waste material, all point to increasing commercial application of crystal chemical principles.

Summary

Diadochy is the limited substitution of ions in a particular mineral species, as well as complete solid solution of two or more ions in related mineral species.

Isomorphism relates a group of minerals of similar morphology and formula that contain different ions in a given coordination site. It does not necessarily imply solid solution within the group.

Isostructural minerals must be identical in all coordination numbers, space group, and number of formulae per unit cell. This does not necessarily imply solid solution.

Limited substitution and complete solid solution are treated as aspects of diadochy, and the term *isomorphism* is abandoned in favor of the term *isostructuralism*.

V. M. Goldschmidt explained substitution by camouflage (i.e., substitution of one element for another of identical charge and similar radius), capture (i.e., substitution of an element with similar radius but higher electric charge), and admission (i.e., substitution of an element with lower charge but similar radius). Admission plus capture balances the electrical charge and results in a coupled substitution.

Where a crystal lattice contains voids that are only partially occupied, defect substitution may result.

Vacancy substitution results when large stray ions or molecules occupy large structural voids.

Crystal field stabilization energy plays a significant role in diadochy.

Substitution of radioactive ions often leads to destruction of the lattice of the host mineral by metamictization, where the daughter product of radioactive decay is much too large to occupy the volume of the site of the original radioactive ion.

Many complete solid solution series are known. Some are limited by requirement of high temperature or high pressure for stability; others are limited by availability of particular elements (e.g., yttrium garnet).

Bibliography

Curtis, C. D. (1964). Application of the crystal field theory to the inclusion of trace transition elements in minerals during magmatic differentiation. *Geochim. et Cosmochim. Acta*, 28: 389–403.

Goldschmidt, V. M. (1937). The principles of distribution of chemical elements in minerals and rocks. *J. Chem. Soc. for 1937, Pt. 1* 655–73.

   (1954). *Geochemistry*, ed. Alex Muir. Oxford at the Clarendon Press, London, pp. 80–125.

Gottfried, D., Jaffe, H. W., and Senftle, F. E. (1959). Evaluation of the lead-alpha (Larsen) method for determining the ages of igneous rocks. U.S. Dept. of Interior, Geol. Survey, Bull. 1097-A, pp. 30–32.

Holland, H. D., and Gottfried, D. (1955). The effect of nuclear radiation on the structure of zircon. *Acta Crystallogr.,* 8: *Pt. 6* 291–300.

Jaffe, H. W. (1951). The role of yttrium and other minor elements in the garnet group. *Am. Mineral.,* 36: 133–55.

 (1955). Precambrian monazite and zircon from the Mountain Pass rare earth district, San Bernardino County, California. *Geol. Soc. Am. Bull.,* 66: 1250.

Yoder, H. S., and Keith, M. L. (1951). Complete substitution of aluminum for silicon: The system $3MnO \cdot Al_2O_3 \cdot 3SiO_2 - 3Y_2O_3 \cdot 5Al_2O_3$. *Am. Mineral.,* 36: 519–33.

10 Density, volume, unit cells, and packing

Density is the concentration of matter measured in mass per unit volume. It is recorded in increments of g/cm³, and is commonly symbolized by the Greek letter ρ. The density of a solid can be obtained by one of the following three methods:

1. $\rho_{(obs)}$ can be determined by direct measurement involving weighing the solid in air and then in a liquid of low density such as toluene, an organic liquid for which density is precisely known and may be corrected for any temperature.

2. $\rho_{(calc)}$ can be determined by calculation from the formula or chemical composition and from the volume of the unit cell measured from X-ray data, using the formula

    $$\rho_{(calc)} = \frac{\text{Formula weight} \times Z \times 1.6603 \times 10^{-24}}{\text{Unit cell volume (Å}^3\text{)} \times 10^{-24}} \quad \text{g/cm}^3$$

    where formula weight is the sum of the mass of its constituent atoms, Z is the number of formulae in one unit cell, and 1.6603×10^{-24} is the reciprocal of Avogadro's number (6.0228×10^{23}). This effectively converts formula weight to gram atom weight, and unit cell volume in Å³ is multiplied by 10^{-24} to convert the volume to cm³.

3. $\rho_{(opt)}$ can be determined by calculation from the optically measured average index of refraction n_{av}, divided by the specific refractivity K of the compound, according to the formula of Gladstone and Dale, where $(n - 1)/K = \rho$ (for details see Chapter 11).

Measured and calculated values of density should agree to within ± 0.1 g/cm³, but they often deviate by larger increments because of experimental error, voids or cavities present in the solid, or fallacious assumptions regarding the chemical composition or purity of the sample. When values are measured and calculated from X-ray data for pure synthetic end members of a solid solution series – for example, pure diopside, $CaMgSi_2O_6$ and hedenbergite, $CaFeSi_2O_6$ – the results should show very close agreement. Densities obtained on chemically analyzed natural mineral solid solutions may be subject to greater error because of the greater chemical complexity of the mineral; for example, small amounts of Na, Al, Ti, Fe^{3+}, or Cr may have been overlooked.

Accurate densities of solids are usually obtained by carefully selecting under the binocular microscope a 25–50-mg pure fragment, then weighing it in air and in toluene, using a sensitive Berman density microbalance. The ratio of these weights, multiplied or corrected by the density of the toluene at the temperature of measurement, gives the density in g/cm^3. Toluene is a good liquid to use because of its low surface tension (27 dyn/cm at 20°C, compared with 73 dyn/cm at 18° for water) and because of its low density compared to that of minerals ($\rho = 0.8867$ at 20°C). Other methods, which involve weighing larger fragments or single crystals on various microbalances or, conversely, weighing fine powders in a pycnometer or suspended in heavy liquids of known density, are all subject to greater experimental error for the reasons already given.

The mineralogist should have some knowledge of the densities of common minerals for use as a yardstick in such calculations. Easy to remember are quartz, 2.65 g/cm^3; magnetite, pyrite, and hematite, near 5.0; native terrestrial or meteoritic Ni-bearing iron, 8.0; and others given in Table 10.1.

Volume

The volume V of a crystal or of any solid is usually obtained from X-ray measurements that yield the dimensions of the unit cell and are most commonly reported in $Å^3$ (less used is the SI unit or nanometer; 1 nm = 10 Å,

Table 10.1. *Range in density of minerals*

Mineral	Range in density (g/cm^3)
Ice	0.91
Hydrated borates	1.4–2.5
Hydrated zeolites	2.0–2.5
Quartz	2.65
Feldspars	2.56–2.78
Ferromagnesian silicates	3.0–4.4
Pyrite, magnetite, hematite	5.0
Other oxides	3.7–7.0
Native elements (nonmetal)	2.0–3.5
Native elements (metal)	7.9–21.5

which results in unit cell dimensions being reported in increments of less than unity).

Unit cell volumes are calculated for minerals crystallizing in the seven crystal systems according to the following formulae (V is in Å^3):

Isometric:
$$V = a^3$$
Tetragonal:
$$V = a^2 c$$
Hexagonal:
$$V = a^2 c \sin 60° \ (\sin 60° = 0.866)$$
Trigonal (rhombohedral):
$$V = (a^3 \sin \alpha) \sqrt{\frac{\cos \alpha - \cos 2\alpha}{1 + \cos \alpha}}$$
Orthorhombic:
$$V = abc$$
Monoclinic:
$$V = abc \sin \wedge \beta$$
Triclinic:
$$V = (abc)\sqrt{1 - \cos^2 \alpha - \cos^2 \beta - \cos^2 \gamma + 2 \cos \alpha \cos \beta \cos \gamma}$$

The reader should understand that the unit cell of a mineral contains one or many formulae of the chemical composition that characterizes the species. Although one unit cell of a simple structure like halite contains only eight atoms (4 Na + 4 Cl), those of more complex structures contain many formulae and atoms; for example, native sulfur contains sixteen formulae of S_8 rings for a total of 128 atoms in one unit cell, and tourmaline contains 162 atoms in the complex unit cell represented by

$$3/NaMg_3Al_6[Si_6O_{18}](BO_3)_3(OH)_4$$

where 3/ means $Z = 3$.

In crystal chemical studies, it is often necessary to distinguish between the number of formulae per unit cell and the number of atoms per unit cell, particularly the number of larger atoms or anions that occupy most of the volume of the cell. Thus the number of oxygen (or O^{2-}) anions per unit cell (Table 10.2) is often cited in studies made of the packing of atoms in response to crystallization under different conditions of temperature and pressure. Because the formula may be cited differently in different texts, the value ascribed to Z may vary, but the number of oxygens or other atoms per unit cell is fixed.

The formula for talc is cited in many texts as $Mg_3Si_4O_{10}(OH)_2$; in other widely used texts it is $Mg_6Si_8O_{20}(OH)_4$ because the doubled formula is more directly comparable to those of amphiboles: talc and mica have $O_{20}(OH)_4$ whereas amphiboles have $O_{22}(OH)_2$, providing a quick recognition of these mineral families from their formulae. Enstatite, given as $MgSiO_3$ in one text and $Mg_2Si_2O_6$ in another, will have $Z = 16$ or $Z = 8$, depending on the formula used, but the total number of oxygens in one unit cell is fixed at forty-eight. An added complication may arise for minerals crystallizing in the trigonal crystal system. Here the unit cell may be delineated by a unit rhombohedron with three equal axes a and an angle between edges α, or by a unit hexagon, each giving different volumes and Z numbers. For example, hematite, Fe_2O_3, has a unit cell described either by $a_{rh} = 5.427$ Å, $\alpha = 55°18'$, with $Z = 2$, or by $a_{hex} = 5.035$ Å and $c_{hex} = 13.749$ Å, with $Z = 6$. The hexagonal unit cell will always have three times the volume of its rhombohedral analogue, and it is useful to know that such minerals tend to translate their unit cells in multiples of three. This can provide valuable information when calculating the density of a mineral. Where Z is not known and the volume and the formula are known, it is usually possible to calculate the density by estimating the value of Z if one has some idea of the density ranges of the common minerals (Table 10.1). To calculate the density of corundum from its formula Al_2O_3 of mass (M) 101.94, the hexagonal unit cell dimensions $a_{hex} = 4.76$ Å and $c_{hex} = 12.996$ Å, and $1/N = 1.6603 \times 10^{-24}$, we use

$$\rho_{(calc)} = \frac{101.94 \times 1.6603 \times 10^{-24}}{V = a^2c \times 0.866 \times 10^{-24}} = 0.6638 \times Z \text{ g/cm}^3$$

If the mineralogist is aware that corundum should be relatively dense, because its lattice is built on a nearly hexagonally close-packed array of O^{2-} anions with four of every six octahedral sites occupied with Al^{3+} ions (from Pauling's Rule 2; see Chapter 5), and that Z is likely to be a multiple of three, then from $\rho_{(calc)} = 0.6683 \times Z$, we would guess that $Z = 3, 6,$ or 9, giving $\rho = 1.99, 3.98,$ or 5.97 g/cm³. It should be obvious that 1.99 is too low, 5.97 is too high, and that 3.98 is the correct value. If the indices of refraction are known or can be

Table 10.2. *Formula units, Z, and oxygen atoms per unit cell*

Mineral	Z	O^{2-} unit cell	Mineral	Z	O^{2-} unit cell
Olivine	4	16	Orthoclase	4	32
Augite	4	24	Albite	4	32
Hypersthene	8 or 16	48	Anorthite	8	64
Tremolite	2 or 4	48	Garnet	8	96
Anthophyllite	4 or 8	96	Leucite	8	96
Biotite	1 or 2	24	Sillimanite	4	20
Kaolinite	1 or 2	18	Andalusite	4	20
Talc	2 or 4	48	Kyanite	4	20
Muscovite	2 or 4	48	Tourmaline	1 or 3	$31_{rh}\ 93_{hex}$
Staurolite	2	48	Beryl	2	36
Analcite	16	112	Hematite	2 or 6	$6_{rh}\ 18_{hex}$
Cordierite	4	72	Melilite	2	14

measured, $\omega = 1.768$ and $\epsilon = 1.760$, then $n_{av} = (2\omega + \epsilon)/3 = 1.7653$, and for various estimates of the specific refractivity k for Al_2O_3 from 0.186 to 0.193, then, from the law of Gladstone and Dale, $(n - 1)/K = \rho$, corundum will have $\rho_{(opt)} = 3.96$–4.11, verifying the choice of $Z = 6$ for the hexagonal unit cell.

Now compare the simple isometric unit cells of kamacite, native iron of meteorites; native gold; diamond; and halite (Fig. 10.1). Note that kamacite has body-centered disposition of its Fe atoms, and, when properly counted as $8 \times (\frac{1}{8})_{cor} = 1, 1 \times 1_{ctr} = 1$, its nine Fe atoms reduce to two per unit cell. Drawn to scale, the structure would show eight Fe atoms in cubic array around every other Fe atom, CN VIII, with all six faces of each unit cube shared. In this orientation, each Fe atom at a cube center would have its $4s^2$ valence electrons overlapped with those of the eight Fe atoms that surround it, resulting in a delocalized metallic orbital that runs through the crystal and explains many of

Figure 10.1. Unit cell projections on (100) locating all atoms in the cubic minerals: native iron or kamacite, native gold, diamond, and halite. Atomic locations and their relationships to mirror planes (m), glide planes ($a, n,$ and d), and screw axes ($4_1, 4_2,$ and 4_3) determine the number of atoms per unit cell, their coordination numbers, the formula, and the space group assignment.

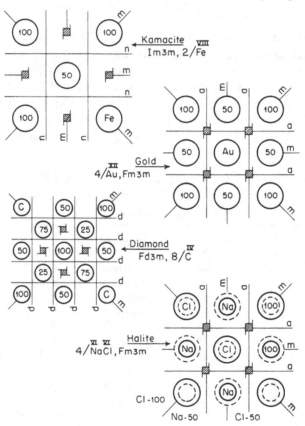

its physical properties. The unit cell formula and the space group are thus written as $2/Fe^{VIII}$ and $Im3m$. Note also that screw axes 4_2 perpendicular to n axial glides accomplish translation of all atoms in a manner identical to or equivalent to that caused by body centering and need not be noted in the space group code; the body centering is, however, a function of the operation of screw axes and glide planes.

Now consider native gold (Fig. 10.1), in which the lattice is face centered with Au atoms located on eight corners and six face centers of the unit cube. When properly counted (see Chapter 6) the fourteen Au atoms reduce to four per unit cell (Fig. 10.1). The array of Au atoms will be in cubic close packing (CCP) in which each Au atom is surrounded by twelve other Au atoms in CN XII. This array is very conducive to the overlap of the spherical $6s^1$ orbitals on each Au atom, and an extended molecular orbital will carry electrons throughout the crystal, in bands, explaining the metallic luster with its high reflectivity, electrical conductivity, and malleability, all characteristic of native gold. The unit cell formula and space group are thus $4/Au^{XII}$ and $Fm3m$.

Also illustrated in Figure 10.1 is the unit cell projection of diamond, in which there are shown eighteen carbon atoms disposed on eight corners, six face centers, and in four of eight cubelets located tetrahedrally at the 75-level (NW and SE) and at the 25-level (NE and SW) inside the unit cell. When the eighteen atoms are properly counted, as explained in Chapters 3 and 6, only eight C atoms belong to the unit cell illustrated. This arrangement and number of atoms results because of disposition of the sp^3 hybrid covalent bonds between all C atoms (see Fig. 3.2).

Note that the three fourfold rotational axes found in kamacite and in native gold are absent from diamond. They are replaced by twelve dextral and sinistral 4_1 and 4_3 screw axes that alternate on and outcrop on each cube face; each such 4_1 and 4_3 axis is perpendicular to diamond glides (d), also four per cubic face (Fig. 10.1). In Figure 3.3, each carbon atom labeled 4, 3, 2, and 1 represents the 100-, 75-, 50-, and 25-elevation levels of the unit cell above the 0 base level $(4 = 100$ and 0); the clockwise (4_1) and counterclockwise (4_3) rotations translate each C atom a net of $\frac{1}{4}$ the distance along the body diagonals of the unit cube. The diamond glides accomplish the same translation of atoms across d glides located at $\frac{1}{8}$, $\frac{3}{8}$, $\frac{5}{8}$, and $\frac{7}{8}$ of the way along each cube face or pair of faces. Reflection across a plane at $\frac{1}{8}$ is accompanied by translation of each C atom, from 100-75-50-25-0 levels (4-3-2-1-0). The unit cell formula and space group are $8/C^{IV}$ and $Fd3m$. The twelve d glides replace the six mirror planes m of kamacite and gold.

The fourth unit cell (see Fig. 10.1) of the familiar halite structure places fourteen chlorine (or sodium) ions on corners and face centers as in native gold, and then adds another thirteen ions of sodium (or chlorine) in the spaces or voids between one another. Both Cl^- and Na^+ are in interchangeable positions with the larger ion, Cl^-, occupying positions of cubic close packing as in native gold. The voids between atomic or ionic populations are all octahedral sites, and all are filled, so six Cl^- ions surround each Na^+ ion, and vice versa. At the same time twelve Cl^- ions surround one another. In halite, all octahedral voids in the CCP array of Cl^- ions are occupied, and all tetrahedral voids are empty. In native gold all such interstitial sites are empty; in kamacite, all voids are tetrahedral, and all are empty and traversed by the 4_2 screw axes. The unit cell of halite contains twice the number of atoms as that of gold, but the symmetry is identical: for halite, $4/Na^{VI}Cl^{VI}$ and $Fm3m$.

All four examples of Figure 10.1 bring atoms close together but use different principal bond types: kamacite and native gold use metallic orbital overlap; diamond, sp^3 covalent orbital overlap; and halite, approximately tangential contact with minimal orbital overlap. The electronegativities of Na ($\Delta\chi = 0.9$) and of Cl ($\Delta\chi = 3.0$) give $\Delta\chi = 2.1$, equivalent to only 66% ionic bonding on Pauling's scale. Thus halite, readily soluble in water and with a low index of refraction, $n = 1.544$, should not be considered a totally ionic compound.

Packing of atoms

The packing index (PI) refers to the degree of efficiency of the occupancy of space by atoms, according to the equation

$$PI \times 10 = \frac{\text{Vol. of ions in 1 unit cell } (\text{Å}^3)}{\text{Vol. of 1 unit cell } (\text{Å}^3)}$$

An "oxygen anion packing index" intended to express the volume of 1 oxygen anion in cm^3 is calculated as follows:

$$cm^3/O^{2-} = \frac{\text{Vol. of 1 mole } V_m(cm^3) \text{ of compound}}{\text{No. of } O^{2-} \text{ anions in 1 mole compound}}$$

In reciprocal relationship, it has been expressed as an "oxygen packing index" or the number of oxygens per cm^3:

$$O^{2-}/cm^3 = \frac{\text{No. of oxygens in 1 mole of compound}}{\text{Vol. of 1 mole, } V_m \text{ (cm}^3\text{) of compound}}$$

Packing indices calculated from these three equations for several selected minerals are compared in Table 10.3. Note that for equal mass of 60.09, the density ρ of stishovite is 4.28, about 1.6 times greater than that of quartz, 2.655 g/cm^3. On this basis, on a scale of 100, 54.9% of space is occupied by O and Si atoms in quartz, and 91.8% in stishovite. Alternatively, the volume of oxygen in 1 mole of quartz is 11.33 cm^3, compared with 7.02 cm^3 in stishovite.

The PI equation requires that atoms be spherical, with a volume $r^3 \times 4\pi/3$, and that orbitals not be overlapped in covalent bonding. It was shown in Chapter 4 that atoms that have their valence electrons in s orbitals form spherical ions, whereas those atoms that place valence electrons in p or d orbitals are spherically symmetrical only if each orbital is singly or doubly occupied (half filled or filled). Thus Fe^{2+}, with six d electrons in five orbitals, and many other

Table 10.3. *Comparison of packing indices calculated for minerals*

Mineral	Mass	/	ρ	=	V_m (cm³)	nO_m	V_m (Å³)	×	Z	=	V_{uc} (Å³)	PI	cm³/O	O/cm³
Anorthite	278.13		2.76		100.77	8	167.31		8		1338.48	5.41[a]	12.60[b]	.0794[b]
Cordierite	585.01		2.505		233.54	18	387.75		4		1551.02	4.94	12.97	.0771
Åkermanite	272.66		2.94		92.74	7	153.98		2		307.96	5.84	13.24	.0755
Calcite	100.09		2.715		36.865	3	61.21		6		367.26	5.83	12.29	.0814
Quartz	60.09		2.655		22.632	2	37.58		3		112.74	5.49	11.32	.0883
Epidote	483.25		3.43		140.89	13	233.93		2		467.85	6.51	10.83	.0923
Forsterite	140.73		3.22		43.70	4	72.56		4		290.26	6.51	10.92	.0916
Fayalite	203.79		4.39		46.42	4	77.08		4		308.30	5.39	11.60	.0862
Diopside	216.58		3.22		67.26	6	111.68		4		446.71	6.54	11.21	.0892
Enstatite	100.41		3.209		31.29	3	51.95		16		831.24	6.97	10.43	.0959
Spinel	142.28		3.55		40.08	4	66.55		8		532.03	6.76	10.02	.0998
Andradite	508.21		3.859		131.69	12	218.66		8		1749.28	6.97	10.97	.0912
Grossular	450.47		3.594		125.33	12	208.11		8		1664.87	7.21	10.44	.0958
Andalusite	162.05		3.145		51.33	5	85.55		4		342.21	6.33	10.31	.0970
Sillimanite	162.05		3.247		49.91	5	82.86		4		331.46	6.55	9.98	.1002
Kyanite	162.05		3.60		45.01	5	74.74		4		298.96	7.28	9.00	.1111
Aragonite	100.09		2.94		34.04	3	56.53		4		226.12	7.06	11.35	.0881
Corundum	101.96		4.00		25.49	3	42.32		6		253.94	7.74	8.50	.1176
Anatase	79.90		3.90		20.49	2	33.97		4		135.88	6.47	10.23	.0978
Brookite	79.90		4.14		19.30	2	32.04		8		256.35	6.87	9.64	.1037
Rutile	79.90		4.29		18.64	2	30.94		2		61.88	7.11	9.39	.1065
Perovskite	135.98		4.03		33.74	3	56.02		1		56.02	8.15	11.25	.0889
Stishovite	60.09		4.28		14.04	2	23.21		2		46.62	9.18	7.02	.1424
Gold	196.97		19.3		10.205	—	16.945		4		67.78	7.41	—	—
Diamond	12.00		3.51		3.419	—	5.676		8		45.41	3.36	—	—
Sphalerite	97.43		4.1		23.76	—	39.46		4		157.82	3.55	—	—

[a] Packing index (PI = V ions$_{uc}$ (Å³)/V_{uc} (Å³) × 10.

[b] Oxygen PI cm³/O = V_m (cm³)/O$_m$ and O/cm³ = 1/(cm³/O).

ions are not spherically symmetrical, and concepts of packing based on spherical ionic volumes are inherently flawed. Indeed, it is the asymmetry of the Fe^{2+} ion that endows it with a preference for distorted rather than symmetrical octahedral sites when both occur in the same mineral. This can be a major factor in cation ordering of asymmetric Fe^{2+} and symmetrical Mg^{2+} ions competing for occupancy of two such sites as in pyroxenes and olivines. Implicit in this concept is the probability that asymmetry of electron density enhances formation of covalent bonds. The closeness of packing of a crate of lemons cannot be equated with that of a crate of oranges.

The other two PI formulae, cm^3/O^{2-} or O^{2-}/cm^3, are based on the assumption that the molar volume (mass/density) of a mineral is a function of the closeness of packing of the oxygen anions alone, and ignores any contribution of large-radius cations. Compare the value obtained for the open-structured framework of quartz, SiO_2, and the densely packed mineral perovskite, $CaTiO_3$:

	PI $\times$ 10	cm^3/O	O/cm^3
Quartz	5.49	11.32	0.0884
Perovskite	8.15	11.25	0.0889

The packing index does represent ideal perovskite as being much more densely packed than quartz, whereas the oxygen packing indices are misleading, in that they represent these two minerals as being identical in packing. In ideal perovskite, large voids in the oxygen-packed structure ($r_{O^{2-}} = 1.40$ Å) are almost exactly filled by large-radius calcium ($r_{Ca^{2+}} = 1.35$ Å). Perovskite represents a cubic close-packed assemblage of large ions, $O^{2-} + Ca^{2+}$, whereas it is an open-packed structure of O^{2-} anions taken alone. Thus, cm^3/O overestimates the volume occupied by one oxygen in a mole of perovskite.

Summary

Density, or concentration of matter measured in mass/unit volume (g/cm^3) is symbolized by ρ. It may be measured directly, calculated from the chemical composition (formula) and X-ray measurement of the unit cell, or calculated from the mean index of refraction n divided by the specific refractivity, K of the compound: $(n-1)/K = \rho$. Density is measured by weighing a pure fragment of the mineral in air and in toluene.

Volume is obtained by X-ray measurement of the dimensions of the unit cell.

The unit cell may contain one or more formulae of the chemical composition of the mineral being measured. For a particular mineral, the number of atoms per unit cell is fixed, whereas the number of formulae per unit cell may vary with the mode of presentation of the formula. In the trigonal system, the true unit cell may be designated by a unit rhombohedron or by a hexagonal cell with a volume three times as large.

Unit cells of four simple isometric minerals are compared and the atoms in them are counted. Kamacite (2 Fe per unit cell) and native gold (4 Au per unit cell) are bonded by overlap of metallic orbitals; diamond (8 C per unit cell) is bonded by sp^3 covalent bonds; and halite (4 Na, 4 Cl per unit cell) is bonded by predominantly ionic bonds.

The packing index, or degree of efficiency of occupancy of space by atoms, may be calculated in three different ways. None of these calculations takes into account all the complexities of crystal chemistry: each method of calculation has a different inherent flaw. Packing indices are calculated by three different methods and tabulated.

Bibliography

Fairbairn, H. W. (1943). Packing in ionic minerals. *Geol. Soc. Am. Bull.*, 54: 1305–74.
Pauling, L. (1960). *The nature of the chemical bond*, 3d ed. Cornell Univ Press, Ithaca, N.Y., chap. 11.

11 Refractivity and polarizability

Optical mineralogy

Optical mineralogy is the study that deals with the identification of minerals from their properties, measured under the polarizing microscope, on crushed grains mounted in liquid reference media. For transparent or nonopaque minerals, with which we will concern ourselves here, the technique basically requires the matching of the index of refraction n of crushed fragments of a mineral with that of a liquid standard whose index of refraction is precisely known to ± 0.002, and can be verified by mounting a drop of the liquid in a refractometer.

All values of n are calibrated for Na light, although measurements are usually made under blue-filtered white light under a microscope fitted with two polarizing prisms (calcite or polaroid) mounted at 90°, one below and one above the rotating object stage of a petrographic microscope. The technique involves the preparation and study of a number of fragment mounts in liquids of different refractive index, until one, two, or all three of the indices of refraction of a particular mineral are determined by matching with the reference liquids. These data, along with a determination of the parallelism or nonparallelism of vibration directions in the crystal with those of the polarizer (lower prism) and analyzer (upper prism) are usually sufficient to identify any transparent mineral. This last statement implies that light vibrates in particular directions in minerals, and, in the general case, it vibrates faster in some directions than in others.

A mineral may have three different indices of refraction – high (γ), intermediate (β), and low (α) – parallel to three mutually perpendicular vibration directions or vibration velocity vectors: Z (slow), Y (intermediate), and X (fast), respectively, if the mineral crystallizes in the triclinic, monoclinic, or orthorhombic crystal systems.

A mineral may have only two vibration velocities – slow, with ϵ parallel to the c axis, and fast, with ω perpendicular to the c axis – for optically positive uniaxial minerals, with these directional velocities reversed for optically negative uniaxial minerals. All uniaxial minerals, whether positive or negative, crystallize in either the tetragonal, hexagonal, or trigonal crystal systems. A mineral will have only one index of refraction n and thus only one vibration velocity constant in all directions if it crystallizes in the isometric crystal system or if it is an amorphous compound.

Now, a bit of data about light, which may be regarded as an electrical disturbance that is part of a spectrum of electromagnetic radiation ranging from short wavelength radiation, X-rays ($\lambda = 1$–10 Å), through ultraviolet radiation, visible radiation ($\lambda = 3900$–7900 Å), to infrared, microwave, and radio-wave radiation of very long wavelength. The velocity of light for this entire electromagnetic spectrum is the same, 3×10^{10} cm/s in a vacuum. An electric vector and a magnetic vector of any such radiation are oriented perpendicular to one another and *vibrate in directions* that are perpendicular, that is, *at right angles to, the direction of propagation of the radiation.*

We will be concerned here with the electric vector, specifically with its vibration direction and velocity in the prism of our microscope and in the minerals we study. Sodium light, our reference standard, vibrates in a vacuum with a wavelength $\lambda = 5893$ Å at the stated velocity $c = 3 \times 10^{10}$ cm/s with a frequency $\nu = 5.09 \times 10^{14}$ cps. These properties in a vacuum are close enough to those in air so that we may treat them as the same. The ratio of these two velocities is close to unity and equal to the index of refraction of air ($n_{air} = c_{vac}/c_{air} = 1.000$). The basic law of the physics of light is $c = \lambda \nu$.

Although light is refracted at most angles of incidence on a crystal plate, at vertical incidence it is retarded but not refracted. When Na light enters the mineral, it is retarded and undergoes a reduction in its wavelength, but its frequency remains unchanged.

If we consider diamond, which has the very high index of refraction $n_{Na} = 2.418$, we can relate the velocity, wavelength, and frequency of Na light in diamond to that in air.

$$\lambda_{Na, dmnd} = \frac{\lambda_{Na, air}}{n_{Na, dmnd}} = \frac{5893 \text{ Å}}{2.418} = 2437 \text{ Å}$$

$$c_{Na, dmnd} = \frac{c_{Na, air}}{n_{Na, dmnd}} = \frac{3 \times 10^{10} \text{ cm/s}}{2.418} = 1.241 \times 10^{10} \text{ cm/s}$$

$$\nu_{Na, air} = \frac{c_{Na, air}}{\lambda_{Na, air}} = \frac{3 \times 10^{10} \text{ cm/s}}{5893 \text{ Å}} = 5.09 \times 10^{14} \text{ cps}$$

$$\nu_{Na, dmnd} = \frac{c_{Na, dmnd}}{\lambda_{Na, dmnd}} = \frac{1.241 \times 10^{10} \text{ cm/s}}{2437 \text{ Å}} = 5.09 \times 10^{14} \text{ cps}$$

To reiterate, then, Na light entering a mineral from air will have its velocity and wavelength reduced and its frequency unchanged.

All minerals of the seven crystal systems fall into one of five optical classes or categories:

1. *Isotropic* or *omniaxial,* in which light vibrates with equal velocity in all directions (isometric minerals)
2. *Anisotropic, uniaxial positive*
3. *Anisotropic, uniaxial negative,* in both of which light vibrates with equal velocity in all directions in one plane perpendicular to the *c* axis, and in which light vibrates with changing velocity in planes parallel to the *c* axis (tetragonal, hexagonal, and trigonal minerals)
4. *Anisotropic biaxial positive*
5. *Anisotropic biaxial negative,* in both of which light vibrates with equal velocity in two planes whose location or orientation is controlled by the magnitude of the differences of the three vibration velocities and resultant indices of refraction (orthorhombic, monoclinic, and triclinic minerals)

A plane in which light vibrates with equal velocity in a 360° rotation is defined optically as a *circular section,* and the direction normal to it is the *optic axis.* From the foregoing discussion, it is evident that isometric minerals have an infinite number of circular sections and an infinite number of optic axes, hence the class, omniaxial.

Tetragonal, hexagonal, and trigonal minerals have but one circular section and, thus, one optic axis: uniaxial positive when the fast velocity and low index of refraction lie parallel to the circular section, and uniaxial negative when the slow velocity and high index of refraction lie parallel to the circular section.

Orthorhombic, monoclinic, and triclinic minerals have two circular sections, and the two optic axes perpendicular to these lie in the plane containing the fast (X) and slow (Z) vibration directions. This X – Z plane is called the *optic plane* (OP) and is usually referred to a prominent crystallographic plane, for example, OP = (010).

Thus, minerals crystallizing in systems based on three unequal crystallographic axes will have three unequal vibration velocities and three indices of refraction, α, β, and γ. Such minerals, containing two optic axes perpendicular to two circular sections, are biaxial positive and biaxial negative. In these minerals, the intermediate velocity Y and intermediate index of refraction β lie parallel to both circular sections. We can model all of these five optical classes by constructing indicatrix models or surfaces that show how the velocity of light changes with direction. By setting each radius of such an indicatrix equal to the one, two, or three indices of refraction of a mineral, we obtain the five indicatrix models shown in Figure 11.1; with $n = r_{\text{indic}}$, all planes become elliptical in the general case, and circular in the special case.

Maximum ellipticity is associated with three indices of refraction that are separated by equal intervals (e.g., $\alpha = 1.600$, $\beta = 1.640$, and $\gamma = 1.680$) and in which circular sections and optic axes both lie at 90° within a biaxial indicatrix.

When the intermediate index of refraction β lies nearer to the low index of refraction α, the ellipticity of the indicatrix is reduced and the mineral is optically biaxial positive (e.g., $\alpha = 1.600$, $\beta = 1.620$, and $\gamma = 1.680$). For an analogous biaxial negative mineral, representative indices are $\alpha = 1.600$, $\beta = 1.660$, and $\gamma = 1.680$. As the β index of refraction moves closer to the γ index of refraction or to the α index of refraction, pairs of biaxial circular sections and the optic axes perpendicular to them approach one another and eventually fuse to produce the uniaxial indicatrix models of Figure 11.1.

In a biaxial positive mineral, it is the α and β indices of refraction that become equal to ω, the low index of refraction and radius of the circular section of a prolate indicatrix (Fig. 11.1); typical indices are $\omega = 1.544$ and $\epsilon = 1.553$, as in quartz. In a biaxial negative mineral these relations are reversed, and β and γ of the biaxial indicatrix become equal to ω, the high index of refraction and radius of an oblate indicatrix (Fig. 11.1); typical indices are $\omega = 1.658$ and $\epsilon = 1.486$, as in calcite.

When a circular section of any mineral lies parallel to the stage of the microscope, the grain will appear dark gray when the polars are crossed (polarizer below and analyzer above the mineral grain), no matter how the stage is rotated. Engaging the convergent lens (located immediately below the stage and the grain) and removing the ocular of the microscope will reveal interference patterns or figures as black maltese crosses, unchanged by stage rotation, for uniaxial minerals and black maltese crosses that split into black crescentic bars

Figure 11.1. Indicatrix models and interference patterns or figures for the five optical classes of crystals using a polarizing microscope with a 50 × objective and a superposed 550 μm gypsum retardation plate.

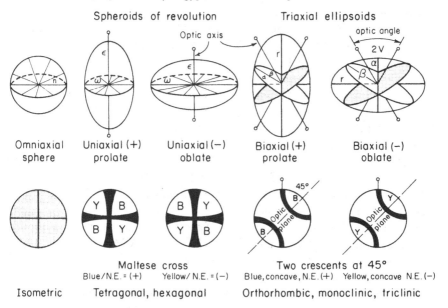

(isogyres) and recombine to form crosses, with stage rotation, in biaxial minerals. Superposition of an accessory gypsum plate into a slot in the microscope tube will add a 550-μm retardation to that produced by the mineral. A net addition or subtraction of wavelength will occur, introducing a blue color into the NE and SW quadrants of the interference figure for positive minerals, or a yellow color into the same quadrants for a negative mineral (Fig. 11.1). Thus, location of the circular section(s) of a mineral, measurement of the associated index of refraction, and observation of the interference figure as indicated become the prime goals of the optical mineralogist. This technique remains the most rapid, most widely used, and best method for the identification of transparent minerals and synthetic compounds.

Refractivity

Over 100 years ago, in 1864, Gladstone and Dale reported that every liquid has a specific refractive energy (specific refractivity) composed of the sum of the specific refractivities of its component elements, modified by the manner of combination, and unaffected by changes in temperature. This specific refractivity accompanies a given liquid when it is mixed with other liquids. The product of this specific refractive energy and its density is, when added to unity, the index of refraction. When the law of Gladstone and Dale, usually expressed in the form $(n - 1)/\rho = K$, is applied to crystalline materials, n is the average or mean index of refraction, ρ is the density, and K is the specific refractive energy of the solid or mineral. Following Larsen and Berman (1921, 1934), k is the specific refractivity of a constituent oxide, p is the weight percent of the constituent oxide, and the sum of the products $k \cdot p = K$ is the specific refractive energy of the mineral; the average index of refraction n_{av} is given as n for isometric minerals or amorphous materials, $(2\omega + \epsilon)/3$ for tetragonal, hexagonal, and trigonal minerals, and $(\alpha + \beta + \gamma)/3$ for orthorhombic, monoclinic, or triclinic minerals, or for omniaxial, uniaxial, and biaxial materials, respectively. Note that an amorphous material such as glass may be regarded as a viscous liquid, turning the full circle between noncrystalline and crystalline materials. These equations used herein for the average index of refraction represent reasonable solutions for the indicatrix models of Figure 11.1 and are more commonly used than $\sqrt[3]{\omega^2\epsilon}$ for uniaxial minerals and $\sqrt[3]{\alpha\beta\gamma}$ for biaxial minerals.

Larsen and Berman (1921, 1934), Tilley (1922), Jaffe (1956), Jaffe, Meyrowitz, and Evans (1953), Jaffe and Molinski (1962), and Mandarino (1976, 1978, 1979, 1981) have applied the law to minerals with considerable success (see also Allen 1956), and requests for reprints of Jaffe's 1956 paper rapidly exceeded his supply. Empirically derived specific refractivity constants k used by Larsen and Berman were modified by Jaffe, then by Mandarino, and again in this work by Jaffe. The new k constants reported herein (Tables 11.1, 11.2) were derived by the author from many mineral and synthetic compounds for which analytical, density, and optical data were available and judged to be reliable. (For comparison, Table 11.3 lists *k* constants determined by Manᵣ darino 1976). Two sets of specific refractivities are presented:

1. Those derived from minerals of a wide spectrum of chemical composition and recommended for general application (Table 11.1)
2. Those derived specifically from rock-forming silicate minerals using the data from the first four volumes of Deer, Howie, and Zussman (1962, 1963 and their abridged 1966 volume) as well as additional data from the recent literature.

Figure 11.2 plots the measured density versus that calculated using the new *k* constants of Table 11.2 for many different silicate mineral types, using chemical composition and optical data published in volumes 1–4 (1962–1963) of Deer et al. Why are two sets necessary? Recall the phrase "modified by the manner of combination" in the original statement of the law by Gladstone and

Figure 11.2. Plot comparing measured density with that calculated from chemical and optical data of Deer, Howie, and Zussman, vols. 1, 2, 3, 4, (1962, 1963).

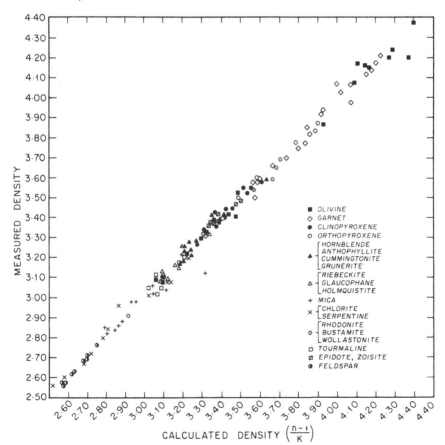

Dale (1864), made before the discovery of X-rays and the advent of modern crystallography. Specific refractive energy values should indeed be expected to vary with the manner of combination of metallic elements with the principal nonmetallic constituent of minerals, oxygen. Here, for the first time, an attempt

Table 11.1. *Relation of specific refractivity k of oxides to valence electron complement and coordination number of cation for mineral components; K values for general use*

Z	V.el.	Comp.	k	C.N.	Z	V.el.	Comp.	k	C.N.
1	$1s^1$	$(NH_4)_2O$	.482	IX	20	$4s^2$	CaO	.226	VI
		H_2O	.340	--				.217	VII
3	$2s^1$	Li_2O	.298	VI				.215	VIII
4	$2s^2$	BeO	.252	IV				.212	IX
5	$2s^22p^1$	B_2O_3	.240	III				.209	XII
			.220	III+IV	21	$4s^23d^1$	Sc_2O_3	.248	VI
			.214	IV	22	$4s^23d^2$	TiO_2	.398	VI
6	$2s^22p^2$	CO_2	.217	III	23	$4s^23d^3$	V_2O_5	.366	IV
7	$2s^22p^3$	N_2O_5	.245	III		$4s^23d^2$	V_2O_4	.277	VI
8	$2s^22p^4$	O	.203			$4s^23d^1$	V_2O_3	.247	VI
9	$2s^22p^5$	F	.043		24	$4s^13d^5$	CrO_3	.330	IV
11	$3s^1$	Na_2O	.178	VIII		$4s^13d^2$	Cr_2O_3	.229	VI
12	$3s^2$	MgO	.241	IV	25	$4s^23d^5$	Mn_2O_7	.335	IV
			.200	VI		$4s^23d^2$	MnO_2	?	opaque
			.201	VIII		$4s^23d^1$	Mn_2O_3	.259	VI
13	$3s^23p^1$	Al_2O_3	.218	IV		$4s^2$	MnO	.288	IV
			.204	IV+VI				.190	VI
			.199	V+VI				.181	VIII
			.187	VI	26	$4s^23d^1$	Fe_2O_3	.332	IV
14	$3s^23p^2$	SiO_2	.206	IV				.280	VI
			.187	VI			(hem.)	.404	VI
15	$3s^23p^3$	P_2O_5	.190	IV		$4s^2$	FeO	.205	IV
16	$3s^23p^4$	SO_3	.177	IV				.185	VI
		S_8	.502	II				.180	VIII
17	$3s^23p^5$	Cl	.303		27	$4s^2$	CoO	.174	VI
19	$4s^1$	K_2O	.186	XII	28	$4s^2$	NiO	.169	VI

is made to correlate specific refractivity with valency, chemical bond type, and coordination number. It appears that, as the coordination of a cation increases, its radius increases (see Table 4.1), the chemical bond becomes more ionic, the electronegativity decreases, the electron density decreases, and the specific refractivity decreases. Conversely, increase in the oxidation state and valency, coupled with low coordination numbers characteristic of covalent bonds, re-

Table 11.1. *(cont.)*

Z	V.el.	Comp.	k	C.N.	Z	V.el.	Comp.	k	C.N.
29	$4s^1 4d^1$	CuO	.172	VI	58	$6s^2 4f^1$	Ce_2O_3	.144	IX
	$4s^1$	Cu_2O	.285	II	59	$6s^2 4f^1$	Pr_2O_3	.142	IX
30	$4s^2$	ZnO	.158	IV	60	$6s^2 4f^1$	Nd_2O_3	.138	IX
			.151	VI	62	$6s^2 4f^1$	Sm_2O_3	.132	IX
31	$4s^2 4p^1$	Ga_2O_3	.159	VI	63	$6s^2 4f^1$	Eu_2O_3	.126	IX
32	$4s^2 4p^2$	GeO_2	.165	IV	64	$6s^2 5d^1$	Gd_2O_3	.120	VIII
33	$4s^2 4p^3$	As_2O_5	.169	IV	65	$6s^2 4f^1$	Tb_2O_3	.116	VIII
	$4s^2 4p^1$	As_2O_3	.199	L&B	66	$6s^2 4f^1$	Dy_2O_3	.111	VIII
34	$4s^2 4p^2$	SeO_2	.147	L&B	67	$6s^2 4f^1$	Ho_2O_3	.106	VIII
35	$4s^2 4p^5$	Br	.214	L&B	68	$6s^2 4f^1$	Er_2O_3	.103	VIII
37	$5s^1$	Rb_2O	.127	XII	69	$6s^2 4f^1$	Tm_2O_3	.101	VIII
38	$5s^2$	SrO	.139	IX	70	$6s^2 4f^1$	Yb^2O_3	.098	VIII
39	$5s^2 4d^1$	Y_2O_3	.168	VIII	71	$6s^2 5d^1$	Lu_2O_3	.095	VIII
40	$5s^2 4d^2$	ZrO_2	.216	VIII	72	$6s^2 5d^2$	HfO_2	.135	VIII
41	$5s^1 4d^4$	Nb_2O_5	.248	VI	73	$6s^2 5d^3$	Ta_2O_5	.138	VI
42	$5s^1 4d^5$	MoO_3	.240	IV	74	$6s^2 5d^4$	WO_3	.140	IV
43-46	Tc - Pd	No data			75	$6s^2 5d^5$	Re_2O_7	.123	?
47	$5s^1$	Ag_2O	.140	VI	76 - 79	Os -Au	No data		
48	$5s^2$	CdO	.121	VI	80	$6s^2$	HgO	.136	VIII
49	$5s^2 5p^1$	In_2O_3	.133	VI	81	$6s^2 6p^1$	Tl_2O	.120	XII
50	$5s^2 5p^2$	SnO_2	.141	VI	82	$6s^2$	PbO	.137	IX
51	$5s^2 5p^3$	Sb_2O_5	.198	?	83	$6s^2 6p^1$	Bi_2O_3	.154	?
52	$5s^2 5p^2$	TeO_2	.199	VI	84 - 89	Po - Ac	No data		
53	$5s^2 5p^5$	I_2O_5	.177	L&B	90	$7s^2 6d^2$	ThO_2	.108	VIII
55	$6s^1$	Cs_2O	.120	XII	91	Pa	No data		
56	$6s^2$	BaO	.125	XII	92	$7s^2 5f^4$	UO_3	.134	II
57	$6s^2 5d^1$	La_2O_3	.149	XII	93	Np	No data		

Table 11.2. *Specific refractivities k for rock-forming silicates*

Oxide	Olivine	Humite	Zircon	Sphene	Garnet	Topaz	Sill.	Andal.	Kyan.	Staur.
SiO_2	.206	.206	.206	.206	.206	.206	.206	.206	.206	.206
B_2O_3	-	-	-	-	-	-	-	-	-	-
Al_2O_3	.186	.193	-	.200	.186	.195	.204	.199	.192	.186
V_2O_3	-	-	-	-	.247	-	-	-	-	-
Cr_2O_3	.250	-	-	-	.245	-	-	-	-	-
Fe_2O_3	.280	.280	-	.275	.280	-	.280	.280	.280	.280
Mn_2O_3	-	-	-	-	-	-	-	-	-	-
TiO_2	.400	.400	-	.392	.450	-	-	-	-	-
ZrO_2	-	-	.200	-	-	-	-	-	-	-
ThO_2	-	-	.112	.110	-	-	-	-	-	-
MgO	.200	.199	-	.200	.201	-	-	-	-	.204
FeO	.191	.191	-	.183	.180	-	-	-	-	.193
MnO	.188	.191	-	.180	.181	-	-	-	-	.192
CaO	.220	.210	-	.212	.216	-	-	-	-	-
NiO	-	-	-	-	-	-	-	-	-	.169
CoO	-	-	-	-	-	-	-	-	-	.172
ZnO	-	-	-	-	-	-	-	-	-	.150
BeO	-	-	-	-	-	-	-	-	-	-
SrO	-	-	-	-	-	-	-	-	-	-
BaO	-	-	-	-	-	-	-	-	-	-
Li_2O	-	-	-	-	-	-	-	-	-	-
Na_2O	-	-	-	.181	.181	-	-	-	-	-
K_2O	-	-	-	.189	.189	-	-	-	-	-
Rb_2O	-	-	-	-	-	-	-	-	-	-
Cs_2O	-	-	-	-	-	-	-	-	-	-
$(NH_4)_2O$	-	-	-	-	-	-	-	-	-	-
Y_2O_3	-	-	.170	.170	.164	-	-	-	-	-
Ce_2O_3	-	-	-	.144	-	-	-	-	-	-
H_2O	.340	.340	.340	.340	.340	.340	.340	.340	.340	.340
F	-	.043	-	-	-	.043	-	-	-	-
O	-	.203	-	-	-	.203	-	-	-	-
Cl	-	-	-	-	-	-	-	-	-	-
CO_2	-	-	-	-	-	-	-	-	-	-

Abbreviations: sill. = sillimanite, andal. = andalusite, kyan. = kyanite, staur. = staurolite, ctoid. = chloritoid, datol. = datolite, epid. = epidote, melil. = melilite, cord. =

Table 11.2. (cont.)

Ctoid.	Datol.	Epid.	Melil.	Beryl	Cord.	Tourm.	Axin.	Opx	Cpx	Amph.	Alk. amph.
.206	.206	.206	.206	.206	.206	.206	.206	.206	.206	.206	.206
–	.213	–	–	–	–	.223	.220	–	–	–	–
									.186jd.		
.186	.193	.190	.229	.193	.213	.193	.198	.200	.200	.208	.186
–	–	–	–	–	–	–	–	–	.230	–	–
–	–	–	–	–	–	–	–	.250	.225	–	–
.280	.280	.280	.280	.280	.280	.280	.280	.265	.265	.280	.270
–	–	.259	–	–	–	–	–	–	–	–	–
.398	–	.400	–	–	–	–	–	.350	.350	.400	.398
–	–	–	–	–	–	–	–	–	–	–	–
–	–	.110	–	–	–	–	–	–	–	–	–
.200	–	.200	.239	.200	.200	.200	.200	.203	.204	.196	.196
.180	–	.185	.215	.180	.180	.185	.185	.183	.205	.182	.180
.187	–	.188	.215	.180	.180	.180	.180	.185	.198	.195	.175
–	.210	.228	.216	.210	.210	.212	.225	.210	.210	.210	.210
–	–	–	–	–	–	–	–	.169	.169	–	–
–	–	–	–	–	–	–	–	–	–	–	–
–	–	–	.156	–	–	–	–	–	–	–	–
–	–	–	–	.260	–	–	–	–	–	–	–
–	–	–	–	–	–	–	–	–	–	–	–
–	–	–	–	–	–	–	–	–	–	–	–
–	–	–	.181	.350	–	.350	–	–	.313	.300	.300
–	–	–	.181	.181	.181	.181	–	.181	.181	.181	.168
–	–	–	–	–	–	–	–	.189	.189	.189	.180
–	–	–	–	–	–	–	–	–	–	–	–
–	–	–	–	.122	–	–	–	–	–	–	–
–	–	–	–	–	–	–	–	–	–	–	–
–	–	–	–	–	–	–	–	–	–	–	–
–	–	.144	–	–	–	–	–	–	–	–	–
–	–	.340	.340	.340	.340	.340	.340	–	–	.340	.340
–	–	–	–	–	–	.043	–	–	–	.043	.043
–	–	–	–	–	–	.203	–	–	–	.203	.203
–	–	–	–	–	–	–	–	–	–	.303	.303
–	–	–	–	–	–	–	–	–	–	–	–

cordierite, tourm. = tourmaline, axin. = axinite, opx. = orthopyroxene, cpx. = clinopyroxene, jd. = jadeite, amph. = Ca, Mg, Fe amphiboles, alk. amph. = alkalic amphiboles, . .

127

Table 11.2. *(cont.)*

Oxide	Pxoid.	Mica	Chlor.	Talc	Felds.	Neph.	Leuc.	Scap.	Analcime	Natrolite
SiO_2	.206	.206	.206	.206	.206	.206	.206	.206	.206	.206
B_2O_3	-	-	-	-	.227	-	-	-	-	-
Al_2O_3	.200	.194	.200	.200	.215	.217	.217	.217	.217	.217
V_2O_3	-	.247	-	-	-	-	-	-	-	-
Cr_2O_3	-	.250	.250	-	-	-	-	-	-	-
Fe_2O_3	.280	.315	.280	.280	.280	-	-	-	-	-
Mn_2O_3	-	-	-	-	-	-	-	-	-	-
TiO_2	.400	.400	.400	-	-	-	-	-	-	-
ZrO_2	-	-	-	-	-	-	-	-	-	-
ThO_2	-	-	-	-	-	-	-	-	-	-
MgO	.204	.204	.196	.204	-	-	-	-	-	-
FeO	.205	.175	.175	.173	.180	-	-	-	-	-
MnO	.190	.171	.187	-	-	-	-	-	-	-
CaO	.223	.210	.210	-	.212	.212	-	.200	-	-
NiO	-	-	-	-	-	-	-	-	-	-
CoO	-	-	-	-	-	-	-	-	-	-
ZnO	-	-	-	-	-	-	-	-	-	-
BeO	-	-	-	-	-	-	-	-	-	-
SrO	-	-	-	-	.135	-	-	-	-	-
BaO	-	.125	-	-	.126	-	-	-	-	-
Li_2O	-	.230	-	-	-	-	-	-	-	-
Na_2O	.181	.181	.181	-	.163	.181	.181	.165	.176	.169
K_2O	.189	.189	-	-	.178	.185	.194	.178	.190	-
Rb_2O	-	.129	-	-	-	-	-	-	-	-
Cs_2O	-	.122	-	-	-	-	-	-	-	-
$(NH_4)_2O$	-	-	-	-	.482	-	-	-	-	-
Y_2O_3	-	-	-	-	-	-	-	-	-	-
Ce_2O_3	-	-	-	-	-	-	-	-	-	-
H_2O	-	.340	.340	.340	-	-	-	.340	.340	.340
F	-	.043	-	-	-	-	-	.043	-	-
O	-	.203	-	-	-	-	-	.203	-	-
Cl	-	.303	-	-	-	-	-	.303	-	-
CO_2	-	-	-	-	-	-	-	.217	-	-
SO_3		-	-	-	-	-	-	.177	-	-

Abbreviations (cont.): . . . pxoid. = pyroxenoid, chlor. = chlorite, felds. = feldspars, neph. = nepheline, leuc. = leucite, scap. = scapolite.

128

Table 11.3. Specific refractivities k derived by Mandarino for general use

Component	Atomic Number	Molecular Weight	k	Remarks	Reliability Indicator
H₂O	1	18.02	0.340		H
Li₂O	3	29.88	0.307		H
(NH₄)₂O	—	52.08	0.483		
Na₂O	11	61.98	0.190		H
K₂O	19	94.20	0.196		H
Cu₂O	29	143.09	0.234		M
Rb₂O	37	186.94	0.128		
Ag₂O	47	231.74	0.168		M
Cs₂O	55	281.81	0.119		
Au₂O	79	409.94	(0.152)		?
HgO	80	417.18	0.144		L
			(0.134)		
Tl₂O	81	424.74	0.115		M-H
Fr₂O	87	462	(0.115)		?
BeO	4	25.01	0.240		H
MgO	12	40.30	0.200		H
			0.225	sulfates & selenates	
SO	16	48.06	0.335		M
CaO	20	56.08	0.210		H
VO	23	66.94	(0.207)		H
CrO	24	68.00	(0.202)		H
MnO	25	70.94	0.197		H
FeO	26	71.85	0.188		H
CoO	27	74.93	0.179		H
NiO	28	74.71	0.176		H
CuO	29	79.55	0.170		H
ZnO	30	81.37	0.158		H
SrO	38	103.62	0.145		H
PdO	46	122.40	0.190		L
CdO	48	128.40	0.130		H
SnO	50	134.69	(0.140)		?
BaO	56	153.34	0.128		H
PtO	78	211.09	0.118		L
HgO	80	216.59	0.123		M
PbO	82	223.19	0.133		M?
RaO	88	242.00	(0.120)		
				sulfates & selenates & nesosilicates & inosilicates	
B₂O₃	5	69.62	0.215		
Cr₂O₃	6	72.02	0.267		?
N₂O₃	7	76.01	(0.325)		
Al₂O₃	13	101.96	0.207		H
			0.242		
			0.176		
P₂O₃	15	109.95	(0.315)		
Sc₂O₃	21	137.91	0.257		H
Tl₂O₃	22	143.80	(0.267)	nesosilicates & inosilicates	H
			(0.227)		H
V₂O₃	23	149.88	(0.279)		H
			(0.237)		H
Cr₂O₃	24	151.99	(0.290)	nesosilicates & inosilicates	M
			(0.247)		M
Mn₂O₃	25	157.87	0.301	nesosilicates & inosilicates	?
			0.256		
Fe₂O₃	26	159.69	0.315	silicates	?
			0.268		M-H
Co₂O₃	27	165.86	(0.329)		?
Ni₂O₃	28	165.42	(0.339)		H
Ga₂O₃	31	187.44	0.170		H
As₂O₃	33	197.84	0.235		H
Y₂O₃	39	225.81	0.170		H
			(0.195)		
In₂O₃	49	277.64	0.130		M?
Sb₂O₃	51	291.50	0.203		M
La₂O₃	57	325.82	0.148		
Ce₂O₃	58	328.24	0.144		H
Pr₂O₃	59	329.81	0.141		H
Nd₂O₃	60	336.48	0.137		H
Pm₂O₃	61	342	(0.133)		?
Sm₂O₃	62	348.70	0.130		H
Eu₂O₃	63	351.92	0.130		H
Gd₂O₃	64	362.50	0.126		H
Tb₂O₃	65	365.85	0.123		H
Dy₂O₃	66	373.00	0.119		H
Ho₂O₃	67	377.86	0.115		H
Er₂O₃	68	382.52	0.112		H
Tm₂O₃	69	385.87	0.108		H
Yb₂O₃	70	394.08	0.104		H
Lu₂O₃	71	397.94	0.101		M
Tl₂O₃	81	456.74	0.097		M
Bi₂O₃	83	465.96	0.053		M
			0.153		
CO₂	6	44.01	0.211		H
SiO₂	14	60.08	0.208		H
PO₂	15	62.97	0.236		
SO₂	16	64.06	0.262		M
TiO₂	22	79.90	0.393		H
VO₂	23	82.94	0.393		H
CrO₂	24	83.99	(0.394)		H
MnO₂	25	86.94	0.394		H
GeO₂	32	104.59	0.167		H
SeO₂	34	110.96	0.195		H
ZrO₂	40	123.22	0.211		H
SnO₂	50	150.69	0.143		H
TeO₂	52	159.60	0.201		M?
CeO₂	58	172.12	(0.205)		L
HfO₂	72	210.49	0.115		
PtO₂	78	227.09	0.151		H
PbO₂	82	239.19	0.105		H
ThO₂	90	264.04	0.167		H
UO₂	92	270.03	(0.100)		?
N₂O₅	7	108.01	0.242		H
P₂O₅	15	141.94	0.183		H
Cl₂O₅	17	150.90	0.220		H
V₂O₅	23	181.88	0.340		H
As₂O₅	33	229.84	0.162		H
Br₂O₅	35	239.81	0.180		
Sb₂O₅	41	265.81	0.268		M?
Nb₂O₅	51	323.50	(0.153)		
Ta₂O₅	73	441.89	0.195		M?
Bi₂O₅	83	497.96	0.151		M?
			(0.139)		
SO₃	16	80.06	0.177		M-H
CrO₃	24	99.99	0.335		
SeO₃	34	126.96	0.164		M?
MoO₃	42	143.94	0.237		M
TeO₃	52	175.60	0.172		M
WO₃	74	231.85	0.171		M
UO₃	92	286.03	0.118		H
S₂O₇	16	176.12	0.133		M
Cl₂O₇	17	182.90	0.182		M
Mn₂O₇	25	221.87	0.348		M
Br₂O₇	35	271.80	(0.156)		?
I₂O₇	53	365.80	0.168		L
Re₂O₇	75	484.40	0.130		
F⁻	9	19.00	0.047		M
Cl⁻	17	35.45	0.318		M
Br⁻	35	79.90	0.217		M
I⁻	53	126.90	0.227		
O²⁻	8	16.00	0.203		H
S²⁻	16	32.06	0.628	sulfur-bearing silicates	H

Constants in brackets were derived by extrapolation or interpolation. Reliability indicators: L (low), M (medium), H (high).

Source: Reprinted with permission from J. A. Mandarino, The Gladstone and Dale relationship, *Can. Mineral.*, 14: 498–502.

sults in a concentration of electron density (overlapped orbitals) and a resulting increase in specific refractivity.

Values for the k constants are derived for oxide constituents of minerals rather than for individual elements or ions because the refractivity of cation and anion or metal and nonmetal are interdependent and specific only for a given coordination. If k values were derived for cations or metals, the values obtained for the nonmetals, principally oxygen anions, would vary all over the map; there is nothing constant about the refractivity of oxygen by itself in minerals. It must be emphasized that the specific refractivity K does not measure the index of refraction; rather, it measures the rates of retardation of the electric vector E of Na light with the mass–volume ratio or density (Fig. 11.3). Values presented here, as well as by others, are averaged from data that show a wide range for some constituents and a narrow range for others, consistent with the factors cited earlier. Note that k values for constituents such as MgO and Al_2O_3 show little variation in different silicate and oxide minerals, whereas those for Fe_2O_3, CuO, and others vary widely between silicate and oxide phases. Evidently this reflects the increased effects of covalent bonding $Fe^{3+} - O^{2-}$ (formal valencies, not ionic charge), and the value for k ($Fe_2^{VI}O_3$) is 0.280 in silicates and a whopping 0.404 in hematite (Table 11.1). The covalency and resulting molecular orbitals permit the refractivity to remain large and undiluted in the oxide, whereas this effect is diluted in the more ionic silicate species.

It has been emphasized that refractivity values, like electronegativity values, are empirically derived and may be expected to vary. Values of K obtained from data on polymorphs of $Si^{IV}O_2^{II}$ are quartz, $K = 0.206$; tridymite, $K = 0.206$; cristobalite, $K = 0.211$; and coesite, $K = 0.203$. The average is the recommended value of 0.206 cited in Table 11.1. Note that for the high pressure polymorph stishovite ($Si^{VI}O_2^{III}$) the increased coordination number causes a decrease of K to 0.187.

A variety of minerals containing tetrahedrally coordinated aluminum give values of $k(Al_2O_3) = 0.215, 0.217, 0.215, 0.220, 0.213, 0.229, 0.227, 0.211$ and an average of 0.218. With coordination number increased to VI, $k(Al_2O_3)$ becomes 0.187 (Table 11.1). The effect of a change in valency or a change in oxidation state on refractivity is very well shown by the oxides of manganese, in which k increases dramatically with valence and is accompanied by a decrease in coordination.

Contrary to popular belief, index of refraction and density need not vary sympathetically, particularly in minerals that contain different principal anions. Minerals of the same or of very similar density may show no correlation with index of refraction (Table 11.4 and Fig. 11.3). When n and ρ are compared for minerals containing one specific anion or for polymorphs or members of isostructural solid solution series, the variation is quite sympathetic. Compare variations of n and ρ for more ionic compounds of F and O with those of the more covalent S, Se, and Te (see Fig. 11.3).

Ionic fluoride minerals have low indices of refraction and relatively high

Figure 11.3. Plot of density (ρ) versus mean index of refraction, $n-1$, for fluorides, oxides, selenides, sulfides, and carbides. Radians are specific refractivity slopes (k) and show that k increases with covalency of anionic component.

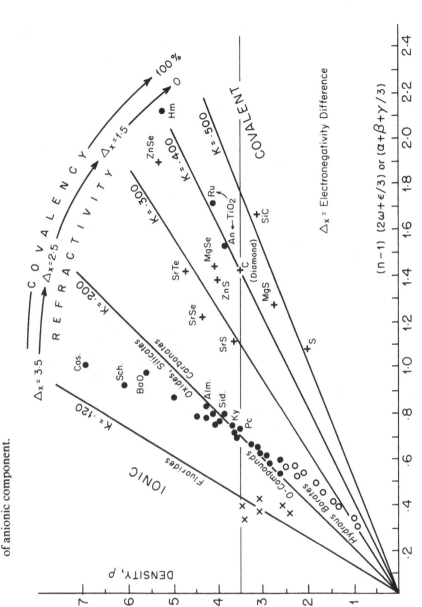

densities, resulting in low specific refractivity values, compared with covalent sulfide and selenide phases that have high indices of refraction and relatively low densities, resulting in high specific refractivity values (Fig. 11.3). Indices of refraction of any species are markedly reduced by the substitution of F for O and/or the substitution of H_2O or (OH) for O; but the low specific refractivity of F ($k = 0.043$) contrasted with the high value for H_2O ($k = 0.340$), results from density contrasts. Note that K-line slopes for solid solution series (Fig. 11.4) increase more rapidly for series containing $Al^{3+}-Fe^{3+}$ than for series containing $Mg^{2+}-Fe^{2+}$. The addition of Fe^{3+} and/or Ti^{4+} to any such mineral will result in a marked increase in the indices of refraction; this expresses the increase in both the number of valency electrons and the covalency.

The practical applications of the law of Gladstone and Dale are many. If n and ρ are measured, it may be used to obtain information on chemical composition. Often the mineralogist has chemical and optical data but cannot measure density, and the simple relation $(n - 1)/K = \rho$ solves the problem in minutes, provided a good table of k values is available. Two such examples cited from Jaffe (1956) and Jaffe et al. (1953) follow.

A spectrographic analysis followed by a preliminary chemical analysis was made of the new mineral sahamalite, $(Mg,Fe)(Ce,La,Nd)_2(CO_3)_4$, named by Jaffe et al. (1953) for the Finnish geochemist Th.G. Sahama. Because only 200 mg of the pure mineral were available, a preliminary chemical analysis was made of a very small quantity, with the following results: $R_2O_3 = 59\%$, $MgO = 6\%$, $FeO = 2\%$, and $CO_2 = 21\%$, for a total of 88 wt.%. After qualitative tests for fluorine and water proved negative, the spectrogram was reexamined. No additional metallic elements could be found. A second spectrogram was made with the same result. By using the measured indices of refraction, the

Table 11.4. *Data showing that index of refraction and density need not vary sympathetically for minerals containing different anions*

Mineral	Formula	Mean refractive index (n)	Density (ρ)
Synthetic	RbBr	1.553	3.41
Synthetic	RbI	1.647	3.48
Synthetic	LiBr	1.784	3.48
Diamond	C	2.412	3.51
Yttrofluorite	$(Ca, Y, \square)F_2$	1.459	3.55
Pyrope	$Mg_3Al_2[SiO_4]_3$	1.714	3.58
Kyanite	$Al_2O[SiO_4]$	1.720	3.60

measured pycnometer density, and the preliminary analytical data, we calculated from $(n = 1)/\rho = K$ that the 12% missing from the analytical summation was a constituent with a specific refractivity k near 0.20. When metallic constituents previously found to be absent by spectrographic analysis were eliminated, this left only CO_2 ($k = 0.217$), Br ($k = 0.214$), I ($k = 0.226$), and O ($k = 0.203$) to account for the missing 12%. The CO_2 analysis seemed to be most suspect and was repeated by two different methods. The correct amount proved to be 31.7% rather than 21%, bringing the analytical summation to 99.+%. The original CO_2 determination was low because the macroabsorption train had recently been set up and was not in equilibrium. Before this was corrected, chemists had argued vehemently that their original CO_2 value could not have been in error!

A second excellent example was provided by a study of the chemically very complex uranium mineral schroeckingerite, described by Jaffe, Sherwood, and Peterson (1948) and Jaffe (1956). Originally, the presence of 2.15% fluorine had gone chemically undetected, leading Larsen and Gonyer (1937) to describe this mineral as a new species they named dakeite. This name was discredited and deleted as a new species when Jaffe et al. (1948) identified the fluorine by spectroscopic analysis and made the necessary chemical analysis. Later, Jaffe (1956) applied the Gladstone and Dale relation to this complex mineral with the excellent result shown in Table 11.5.

Earlier, it was noted that different coordination numbers for a cation would result in different refractivity values. Differences in coordination number, and,

Figure 11.4. Increase in refractivity in isostructural series. Note that k-line slopes increase rapidly (they are flatter) for series with Al^{3+}—Fe^{3+} than for those with Mg^{2+}—Fe^{2+}.

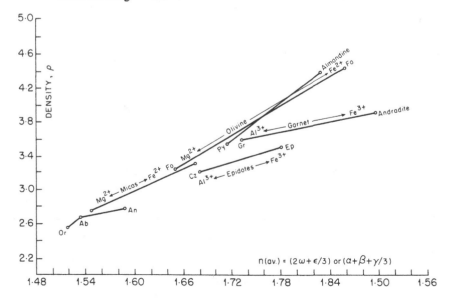

indeed, differences in distortion of a site with the same formal coordination number may be expected to influence a given refractivity value.

Polarizability

Closely related to refractivity is the phenomenon of electronic polarizability, which refers to the temporary displacement of valence electrons in an atom or ion induced by the electric vector of electromagnetic radiation operative at optical frequencies (v_{Na} = 5.09 × 10^{14} cps). After this type of displacement, centers of gravity of the atomic nucleus and the electric charge no longer coincide, and the atom acquires an induced dipole moment μ. The magnitude of such dipole moments summed over all the atoms in a mineral or molecule can be evaluated either by (1) calculation of the molecular electronic polarizability α ($Å^3$), from measurements of indices of refraction, density, and molar volume, or from (2) measurement of a dielectric constant ϵ. At optical frequency the mean index of refraction n and dielectric constant ϵ are related by $\epsilon = n^2$ or $\sqrt{\epsilon} = n$, as discovered independently by Lorentz and Lorenz in 1880.

Here, an analogy with a marathon runner is appropriate. The runner, moving across a meadow with a speed of 9 mph, is slowed to 6 mph on entering and passing through a spruce forest (Fig. 11.5). Branches of the spruce trees are momentarily elastically displaced by the runner in a manner analogous to the

Table 11.5. *Comparison of measured and calculated mean indices of refraction from (α + β + γ)/3 for schroeckingerite, a chemically very complex uranium mineral*

Oxide	$p/100$		k		$(k)(p/100)$	
CaO	18.93	×	.225	=	.04259	
Na_2O	3.49	×	.181	=	.00632	
UO_3	32.19	×	.134	=	.04313	
CO_2	14.86	×	.217	=	.03225	
SO_3	9.01	×	.177	=	.01595	
H_2O	20.28	×	.340	=	.06895	
F	2.14	×	.043	=	.00092	
	100.90				.21011	
—O=F	.90	×	.203	=	.00183	
	100.00		K	=	.20828	α = 1.489
			d	=	2.51	β = 1.542
			dK	=	.52278	γ = 1.542

Note: Mean index, $dk + 1$ = 1.523 (n calculated). Mean index, (α + β + γ)/3 = 1.524 (n experimental).

Source: Reprinted with permission from H. W. Jaffe (1956), Application of the rule of Gladstone and Dale to minerals, *Am. Mineral.*, 41: 759.

momentary elastic displacement of electrons in an atom by the electric vector of electromagnetic radiation, or visible light. The thicker the forest, the greater the reduction in the runner's velocity, hence, the greater his or her retardation. Similarly, the higher the electron density and covalency in a mineral, the greater will be the electronic polarizability of its atoms (Fig. 11.5).

Polarizability should increase with numbers of valence electrons affected and by the length of orbitals of atoms of different radii. Within a vertical group of the periodic table, polarizability should increase downward together with both volume and mass in an opposite sense to electronegativity and ionization potential values.

Equations relating specific refractivity K, molecular electronic polarizability

Figure 11.5. Relation between polarizability, retardation, and chemical bonding, analogous to a marathon runner entering a spruce forest. Refractivity, K, polarizability, $\alpha_{G(\text{Å}^3)}/V_{m(\text{Å}^3)}$, and index of refraction, n, increase with covalent overlap in the sequence: MgF–MgO–MgS.

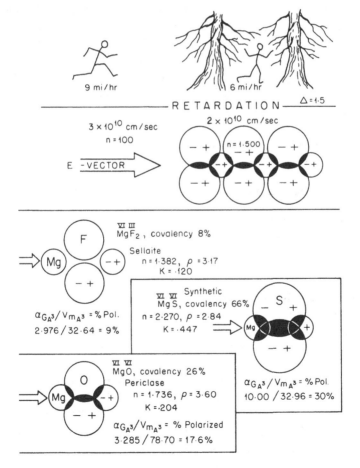

α, and molecular refractivity R to the time-honored formulae of Lorentz–Lorenz and Gladstone and Dale are as follows:

Lorentz–Lorenz $\dfrac{\epsilon - 1}{\epsilon + 2} = \dfrac{n^2 - 1}{n^2 + 2} \cdot \dfrac{1}{\rho} = K_L$ (1)

Gladstone–Dale $\dfrac{n^2 - 1}{n + 1} \cdot \dfrac{1}{\rho} = \dfrac{n - 1}{\rho} = K_G$ (2)

$$\frac{K \cdot M}{4\pi/3N} = \frac{R\,(\text{cm}^3)}{2.523} = \alpha_L\,(\text{Å}^3) \quad \text{or} \quad \alpha_G\,(\text{Å}^3)$$ (3)

depending on whether K_L or K_G is used,

$$\frac{n - 1 \cdot V_m\,(\text{Å}^3)}{4\pi/3} = \frac{R\,(\text{Å}^3)}{4.189} = \alpha_G\,(\text{Å}^3)$$ (4)

where K_L = specific refractivity of Lorentz and Lorenz, K_G = specific refractivity of Gladstone and Dale, ϵ = the dielectric constant measured at v_{Na} = 5.09×10^{14} cps, n = the mean index of refraction measured for Na light, λ = 5893 Å, M = mass, V_m = molar volume (in cm^3 or Å^3), R = molar or molecular refractivity (in cm^3 or Å^3), N = Avogadro's number = 6.0228×10^{23} g/mole, $4\pi/3$ is a Lorentz function or factor describing the mutual interaction of atomic dipoles, ρ = density in g/cm^3, α_L and α_G (Å^3) = molecular electronic polarizability at optical frequency. Note that $4\pi/3 = 4.189$ and $4\pi/3N = 2.523$.

Now $K_L, K_G, \alpha_L, \alpha_G, R_L$, and R_G may be calculated from equations (1), (2), (3), and (4) from the following data for α-quartz, SiO$_2$:

$$n_\omega = 1.544, \quad n_\epsilon = 1.553, \quad n = 1.547$$

$$M = 60.09/\rho = 2.655 = V_m = 22.632 \text{ cm}^3 \text{ or } 37.575 \text{ Å}^3$$

These data give for α-quartz:

$$K_L = 0.1194 \quad \alpha_L = 2.843 \text{ Å}^3 \quad R_L = 7.174 \text{ cm}^3 \ (11.91 \text{ Å}^3)$$

$$K_G = 0.206 \quad \alpha_G = 4.907 \text{ Å}^3 \quad R_G = 12.378 \text{ cm}^3 \ (20.551 \text{ Å}^3)$$

In this book we use specific refractivity values K and molecular electronic polarizability values α_G (Å^3) derived from the Gladstone–Dale equation. If the Lorentz–Lorenz equation is used, the calculated polarizability values α_L (Å^3) will yield values comparable to those reported by Lasaga and Cygan as optical polarizabilities.

A list of some of the molecular electronic polarizabilities α_G (Å^3), derived from selected minerals in this study, is presented in Table 11.6. Just as specific refractivity K does not measure index of refraction, neither does the absolute value of polarizability α_G measure index of refraction. What matters is the *polarizability in a given molar volume.*

Table 11.6. *Polarizability of minerals: % mol polarized and relation of the %
molar volume polarized to the mean index of refraction*

Mineral	α_G (Å³)	/ V_m (Å³) = %	· 4π/3	= (n − 1)	/ ρ	= K
Villiaumite	1.956	24.98	.0783	4.189	.328	2.79	.1175
H₂O	2.378	29.91	.0795		.333	1.00	.340
Sellaite	2.976	32.64	.0912		.382	3.17	.1205
Fluorite	4.213	40.76	.1034		.433	3.18	.136
Borax	25.615	231.76	.1105		.463	1.70	.272
Albite	21.05	166.12	.1267		.531	2.62	.2027
Halite	5.83	44.926	.1298		.544	2.16	.252
Quartz	4.907	37.58	.1306		.547	2.655	.206
Anorthite	23.33	167.36	.1394		.584	2.76	.2116
Coesite	4.88	34.17	.1428		.598	2.92	.205
Calcite	8.81	61.32	.1437		.601	2.71	.222
Aragonite	8.49	56.33	.1507		.632	2.95	.214
Andalusite	12.93	85.55	.1511		.633	3.145	.201
Magnesite	7.13	46.98	.1518		.636	2.98	.2134
MgCl₂	10.53	68.15	.1545		.647	2.32	.279
Forsterite	11.27	72.56	.1553		.652	3.22	.202
Sillimanite	13.10	82.26	.1581		.662	3.247	.2042
Pyrope	31.84	186.88	.1704		.714	3.582	.1993
Kyanite	12.62	73.41	.1719		.720	3.665	.1965
Grossular	36.42	208.91	.1743		.734	3.594	.2042
Periclase	3.285	18.70	.1757		.736	3.58	.2056
Rhodochrosite	9.15	51.58	.1774		.743	3.70	.2008
Corundum	7.74	42.32	.1829		.766	4.00	.1915
Siderite	9.22	48.58	.1898		.795	3.96	.2008
Spessartine	37.46	196.15	.1910		.800	4.19	.1909
Stishovite	4.46	22.93	.1945		.815	4.35	.1874
Almandine	37.92	191.40	.1981		.830	4.318	.1922
Lime	5.65	27.63	.2045		.838	3.37	.254
SrO	7.11	34.14	.2083		.870	5.04	.173
Andradite	46.29	218.65	.2117		.887	3.859	.2298
BaO	10.41	44.51	.2339		.980	5.72	.1713
Cassiterite	7.86	31.99	.2457		1.029	6.99	.1472
Sulfur	52.67	204.27	.2578		1.080	2.085	.518
MgS	10.00	32.96	.3033		1.271	2.84	.4475
Sphalerite	12.90	39.46	.3269		1.370	4.1	.334
Diamond	1.913	5.676	.3370		1.412	3.51	.4023
Anatase	12.48	34.01	.3670		1.537	3.90	.394
Brookite	12.41	32.04	.3873		1.622	4.14	.392
Rutile	12.60	30.99	.4066		1.708	4.29	.398

The significant relation is

$$\alpha_G/V_m \, (\text{Å}^3) \times 4\pi/3 = n - 1 \tag{5}$$

For example, borax, $Na_2B_4O_5(OH)_4 \cdot 8H_2O$, has the high Gladstone–Dale polarizability 25.615 Å^3 but the large molar volume 231.76 Å^3; equation (5) yields $n = 1.463$, a low index of refraction. Diamond has a low α_G of 1.913 Å^3, but its very small molar volume, 5.67 Å^3, yields $n = 2.412$ from equation (5) (Table 11.6).

Table 11.7 lists molecular polarizabilities α_G derived for oxides that form their own minerals and that are components of more complex silicate minerals. Table 11.8 illustrates the additivity of several of these oxide polarizabilities to obtain those of more complex silicate minerals, such as kyanite, forsterite,

Table 11.7. *Polarizability of some oxides in minerals*

Oxide & CN	α_G	Mineral example	Oxide & CN	α_G	Mineral example
$Si^{IV}O_2$	4.907	Quartz	$K_2^{IX}O$	7.206	Sanidine
$Al_2^{VI}O_3$	7.76	Corundum	$Fe^{VI}O$	5.346	Siderite
$Al_2^{IV}O_3$	8.26	Sillimanite	$Mn^{VI}O$	4.926	Rhodochrosite
$Mg^{VI}O$	3.28	Periclase	$Fe_2^{VI}O_3$	16.82	Andradite
$Ca^{VI}O$	4.925	Calcite	$Ti^{VI}O_2$	12.615	Rutile
$Ca^{IX}O$	4.655	Aragonite	$Sr^{VI}O$	7.092	SrO
CO_2	3.872	Magnesite	$Ba^{VI}O$	9.89	BaO
$Na_2^{VIII}O$	4.566	Jadeite	$Sn^{VI}O_2$	8.80	Cassiterite
H_2O	2.375	H_2O	(SO_3)	6.11	Anhydrite

Table 11.8. *Some examples of the additivity of polarizabilities of oxides to yield values for silicates*

	α_G		α_G
Corundum	7.76	3 periclase	9.846
+	+	+	+
Quartz	4.907	1 corundum	7.76
"kyanite"	12.667	+	+
Kyanite	12.62	3 quartz	14.721
		"pyrope"	32.327
		Pyrope	32.06
Quartz	4.907		
+	+		
2 periclase	6.56		
"forsterite"	11.467		
Forsterite	11.292		

pyrope, and anorthite. The agreement between the polarizabilities α_G of these minerals and the sums of their constituent oxides is indeed good. The reader will profit from designing some additional examples from the data of Tables 11.7 and 11.8

Electronic polarizabilities have been calculated for several ions by Fajans and Joos (1924), Born and Heisenberg (1924), Pauling (1927), Tessman, Kahn, and Shockley (1953), and Lasaga and Cygan (1982) (Table 11.9). These polarizabilities were determined from the Lorentz–Lorenz equation, and although the numbers obtained will differ from these presented here, the results obtained from each formula are internally consistent.

Optical geochemistry

Perhaps the ultimate application of data on refractivity and polarizability lies in the correlation of the optical properties of minerals with their

Table 11.9. *Electronic polarizabilities of ions in Å³*

Ion	Tessman, Kahn, & Shockley (1953)	Pauling (1927)	Born & Heisenberg (1924)	Fajans & Joos (1924)	Lasaga & Cygan (1982)
Li^+	.03	.029	.075	.08	.03
Na^+	.41	.179	.21	.196	1.14
K^+	1.33	.83	.87	.88	1.98
Be^{2+}	—	.008	—	.04	.05
Mg^{2+}	—	.094	.012	.12	.48
Ca^{2+}	1.1	.47	—	.51	1.66
Sr^{2+}	1.6	.86	1.42	.86	—
Ba^{2+}	2.5	1.55	—	1.68	—
B^{3+}	—	.003	—	.02	—
Al^{3+}	—	.052	.065	.067	.13
La^{3+}	—	1.04	—	1.3	—
Si^{4+}	—	.016	.043	.04	.08
Ti^{4+}	—	.185	—	.236	—
O^{2-}	.5–3.2	3.88	—	2.75	1.31
S^{2-}	4.8–5.9	10.2	—	8.6	—
Se^{2-}	6.0–7.5	10.5	—	11.2	—
Te^{2-}	8.3–10.2	14.0	—	15.7	—
F^-	.64	1.04	.99	.98	—
Cl^-	2.96	3.66	3.05	3.53	—
Br^-	4.16	4.77	4.17	4.97	—
I^-	6.43	7.10	6.28	7.55	—
$[CO_3]^{2-}$	2.7–5.2	—	—	—	—
$[SO_4]^{2-}$	3.0–5.3	—	—	—	—

crystal chemistry. This new field has been named *optical geochemistry* by this author, who teaches the discipline as a graduate course at the University of Massachusetts.

Bragg (1933) showed that optical character or class and birefringence, B (the difference between the maximum and minimum indices of refraction for constant thickness), are related to crystal structure. According to Bragg, planar structures such as calcite (see Fig. 19.6) or sassolite (see Fig. 17.2), with all planar (CO_3) or (BO_3) groups lying *perpendicular* to the c crystal axis, are optically negative with high birefringence (Fig. 11.6). Bastnaesite (see Fig. 19.8), $(Ce,La,Nd)(CO_3)F$, is optically positive because the planar (CO_3) groups are oriented *parallel* to the c axis. Bragg also noted that atoms lying in rows staggered parallel to the c crystal axis will have optically positive character and high birefringence, for example, rutile (Fig. 11.7; also see Fig. 18.9).

These relations have been studied extensively by this author, who has his graduate students build packing models of minerals from the atomic coordinate crystal structure maps known as unit cell projections. The projections and the models can be used to decipher and relate the travel paths of different velocities of light and the magnitude of their differences in many minerals. A summary of these relations follows.

Figure 11.6. Minerals built of layers perpendicular to the c axis are optically negative with high birefringence. Polarizability is high parallel to the layers, low perpendicular to the layers in minerals with $[CO_3]^{2-}$, $[NO_3]^-$, and $[BO_3]^{3-}$ groups oriented perpendicular to the c axis.

VIEW DOWN C-AXIS

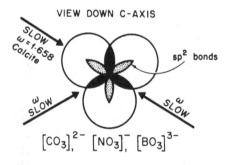

$$\left[CO_3\right]^{2-} \left[NO_3\right]^- \left[BO_3\right]^{3-}$$

VIEW // C-AXIS

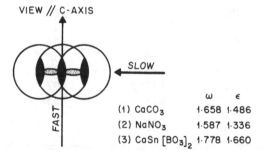

	ω	ϵ
(1) $CaCO_3$	1·658	1·486
(2) $NaNO_3$	1·587	1·336
(3) $CaSn[BO_3]_2$	1·778	1·660

It will be apparent that structural geologic elements (foliation and lineation of strata) may be used as analogues of crystal chemical elements (planar atom and linear atom arrays). Minerals, like rocks, may be regarded as foliated, lineated, or massive, with respect to the degree of anisotropy of electron density conferred by chemical bonding and crystal structure.

Minerals with maximum electron density paths oriented in layers perpendicular to the c axis (phyllosilicates, carbonates, nitrates) will be optically negative with a high birefringence B. In calcite and soda-niter, covalent hybrid bond sp^2 orbitals impose short $O-O$ distances of 2.22 Å and 2.15 Å in planar (CO_3) and (NO_3) groups, conferring high negative B of 0.172 and 0.251, respectively.

Similarly, talc, muscovite, and pyrophyllite are optically negative with high B. Chlorite, however, though equally well layered perpendicular to the c axis, may be positive or negative around the optic angle (Fig. 11.1), $2V = \pm 0°$ because of the H^+ ion or proton. Brucite and gibbsite are positive with moderately high B because the H^+ bonds extensively polarize the O ions parallel to the c axis. When the brucite-like layer is combined with an aluminous talc-like layer in chlorite, interlayer tetrahedral–octahedral $O-H-O$ hydrogen bonds parallel to the c axis (optically positive) neutralize the (Fe, Mg, Al)$-O$ Si$-O$ layers (optically negative) to yield low B and 2V near 0°. With Fe predominant, the octahedral "talc" layer has electron density slightly greater than that of the

Figure 11.7. Minerals with shared-edge-shortened polyhedra lying in staggered rows parallel to the c axis will be optically positive with high birefringence, as in rutile, TiO_2. High electron density is linear parallel to c, and not in layers. Zircon and anhydrite are similar.

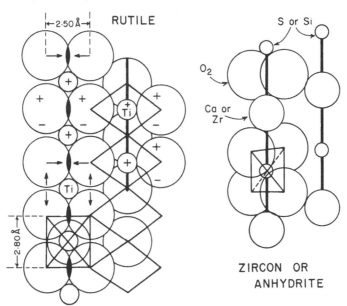

RUTILE

|←2·50 Å→|

S or Si

O_2

Ca or Zr

Ti

Ti

|←2·80 Å→|

ZIRCON OR ANHYDRITE

"brucite" layer, and chlorite is optically negative and length slow; with Mg dominant, chlorite is optically positive and length fast (Fig. 11.8).

If atoms and electron density maxima are oriented in staggered rows or spindles parallel to the c axis (rutile, anhydrite, zircon), high optically positive B will result. In rutile, opposed shared-edge shortening of octahedral edges forms ladders parallel to the c axis (lineation) to yield a very high ϵ of 2.901, whereas light vibrating parallel to the a axes traverses discontinuous octahedral sites interrupted by voids, with a much lower electron density of $\omega = 2.605$ and $B = 0.296$. In anatase, each octahedron shares four edges in two pairs oriented at right angles to one another, both lying in planes parallel to the a axes. In the c axis direction, every other octahedral site is empty. Along a, the pairs of shared-edge-shortened $O-O$ octahedral edges zigzag or undulate to build up continuous high electron density paths, contrasting with void-interrupted low electron density paths parallel to c. Anatase is optically negative with $B = 0.073$ (see Figs. 18.11, 18.12).

Anhydrite is strongly layered with O, Si, and Ca atoms lying parallel to the optic plane (100) and with all (SO_4) tetrahedra lying within intersecting (011) layers. Yet anhydrite is optically positive because opposed keels of (SO_4) tetra-

Figure 11.8. In sheet silicates, O^{2-} in tetrahedral and octahedral layers perpendicular to c is strongly polarized perpendicular to c, but interlayer H^+ strongly polarizes O parallel to c. Thus H-bonds parallel to c neutralize or counter high polarizability of O^{2-} layers perpendicular to c. Layer lattices are normally optically negative, but H bonds can reverse optic sign to positive. In upper series of drawings, octahedral layers (in rectangles) lie between tetrahedral Si_4O_{10} layers. Lower series of drawings shows decreasing ellipticity of optical indicatrix planes and wave velocity contrasts expressed as index of refraction contrasts.

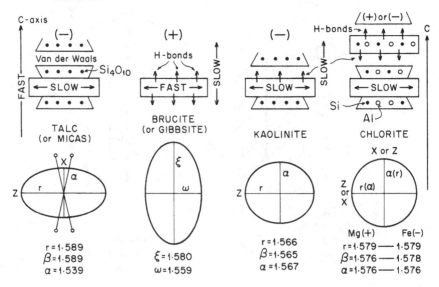

hedra alternate with Ca ions in staggered rows or spindles of high electron density parallel to the c axis. These keels form infinite chains of $Ca-O_2-S-O_2$. In the (100) optic plane, $Z = c$, $X = b$, and $r - \alpha = 1.609 - 1.569 = B = 0.040$. The uniaxial positive zircon structure is similar, with opposed (SiO_4) keels forming, with Zr, infinite spindles of high electron density of $Zr-O_2-Si-O_2$ parallel to c (Figs. 11.7; also see Figs. 12.5, 19.1).

Minerals with equidimensional polyhedral groups, such as tetrahedra, sharing corners, as in tektosilicates (feldspars, feldspathoids, zeolites) are predominantly ionic and do not build up paths of high electron density. These minerals are either isotropic or have low birefringence. In an electric field, bonded electrons cause F, O, and S ions and atoms to behave as a billiard ball, a tennis ball, and a balloon, respectively, resulting in corresponding increases in polarizability, refractivity, and mean index of refraction. The sequence F, O, S is one of decreasing electronegativity and increasing covalency, electron density, and refractivity (Fig. 11.3).

In addition to the effects of orientation of shared-edge-shortened packages, planes, and rows, the effect of the orientation of periodic bond chains (Hartman and Perdok 1955) or unbroken links and chains may be a principal cause of high polarizability in some minerals. For example, in sphene (see Fig. 12.3), the kinked bond string of $O-Ti-O-Ti---$ correlates with the orientation of the slow vibration direction and maximum index of refraction.

In the descriptive section that follows (Chapters 12–20), crystal chemical analyses of selected minerals are made, which illustrate many of the principles put forth in this chapter and earlier in this book.

Summary

Optical mineralogy is the best single method for identifying transparent minerals. It involves

1. The measurement under the microscope of one, two, or three principal indices of refraction by matching these with calibrated reference liquids
2. A determination of the optic sign and optical class of a given mineral. There are five optical classes to which all minerals may be assigned, depending on the number of circular sections (CS) or equal velocity planes they contain, and their orientation inside the crystal.

An optic axis (OA) emerges perpendicularly from each circular section. The five optical classes are

1. Omniaxial (CS = ∞; hence OA = ∞)
2. Uniaxial positive and
3. Uniaxial negative (each with one CS and one OA)
4. Biaxial positive and
5. Biaxial negative (each with two CS and two OA).

These classes may be modeled as ellipsoids of revolution from radii equal to indices of refraction, as a

1. Spheroid
2. Prolate spheroid of revolution
3. Oblate spheroid of revolution
4. Triaxial ellipsoid (prolate)
5. Triaxial ellipsoid (oblate)

All isometric minerals and amorphous compounds crystallize in Class 1, tetragonal, hexagonal, and trigonal minerals in Classes 2 or 3, and all orthorhombic, monoclinic, and triclinic minerals in Classes 4 or 5.

The electric vector of electromagnetic radiation of Na or visible light (E_{Na} or E_{vis}) is retarded and, in most cases, refracted on entering a crystal plate or grain from the air; thus, the velocity c is reduced, the wavelength λ is also reduced, but the frequency v remains unchanged.

Refractivity and polarizability are phenomena that express the retardation of the E vector of light by a mineral of given composition, mass, molar volume, and density.

Specific refractivity $K = (n - 1)/\rho$ measures the rate of retardation of E_{Na} with density.

Molecular polarizability α_G measures the total of electric dipole moments operative in a crystal, and α_G/V_m (Å^3) $\times 4\pi/3 = n - 1$. Thus, the polarizability in a given molar volume of mutually interacting dipoles is the index of refraction, just as is the product of the specific refractivity and density $+$ unity. One may be derived from the other.

Both α_G and K are reasonably additive when oxide values are combined to sum to complex silicate values if reasonable attention is paid to choosing the correct value for a given CN of a component. New and revised lists of K and α_G constants or values for use in mineralogical calculations are presented.

The relationships between the vibration directions or vibration velocity vectors (Z = slow, Y = intermediate, X = fast) and their orientation in crystals depends essentially upon the arrays of atoms or ions. Thus, refractivity and crystal chemistry (optical geochemistry) form a rewarding, if somewhat neglected, subject for added study.

Bibliography

Ahrens, L. H. (1959). Variation of refractive index with ionization potential in some isostructural crystals. *Min. Mag.* 31: 929.

Allen, R. D. (1956). A new equation relating index of refraction and specific gravity. *Am. Mineral.*, 41: 245–57.

Batsanov, S. S. (1961). *Refractometry and chemical structure.* Consultants Bureau, New York.

Bloss, F. D. (1961). *An introduction to the methods of optical crystallography.* Holt, Rinehart and Winston, New York.

Born, M., and Heisenberg, K. (1924). Uber den Einfluss der Deformierbarkeit der Ionen auf optische und chemische Konstanten. I. *Z. Phys.,* 23: 388–410.

Bragg, W. L. (1933). *The crystalline state.* Vol. I. *A general survey.* G. Bell, London.

Cygan, R. T., and Lasaga, A. C. (1986). Dielectric and polarization behavior of forsterite at elevated temperatures. *Am. Mineral.,* 71: 758–66.

Dana, E. S., and Ford, W. E. (1932). *A textbook of mineralogy,* 4th. ed. Wiley, New York.

Deer, W. A., Howie, R. A., Zussman, J. (1962–1963). *Rock-forming minerals,* Vols. 1–5. Longman, London.

(1966). *Introduction to the rock-forming minerals.* Longman, London.

Fajans, K., and Joos, G. (1924). Molrefraktion von Ionen und Molekülen im Lichte der Atomstruktur. *Z. Phys.,* 23: 1–46.

Gladstone, J. H., and Dale, T. P. (1864). Researches on the refraction, dispersion, and sensitiveness of liquids. *Roy. Soc. London Philos. Trans.,* 153: 337.

Hartman, P., and Perdok, W. G. (1955). On the relations between structure and morphology of crystals – I. *Acta Crystallogr. Pt. 1* 8: 49–52.

Jaffe, H. W. (1956). Application of the rule of Gladstone and Dale to minerals. *Am. Mineral.,* 41: 757–77.

Jaffe, H. W., Meyrowitz, R., and Evans, H. T. (1953). Sahamalite, a new rare earth carbonate mineral. *Am. Mineral.,* 38: 749.

Jaffe, H. W. and Molinski, V. J. (1962). Spencite, the yttrium analogue of tritomite from Sussex County, New Jersey. *Am. Mineral.* 47: 9–25.

Jaffe, H. W., Sherwood, A. M., and Peterson, M. J. (1948). New data on schroeckingerite. *Am. Mineral.,* 33: 152–7.

Larsen, E. S., Jr., and Berman, H. (1934). The microscopic determination of the nonopaque minerals. U.S. Dept. of Interior, *Geol. Surv. Bull.* 848, 2d ed. (1921, 1st ed.).

Larsen, E. S., Jr., and Gonyer, F. A. (1937). Dakeite, a new uranium mineral from Wyoming. *Am. Mineral.,* 22: 561–3.

Lasaga, A. C., and Cygan, R. T. (1982). Electronic and ionic polarizabilities of silicate minerals. *Am. Mineral.,* 67: 328–34.

Lorentz, H. A. (1880). Ueber die Beziehung zwischen der Fortpflanzungsgeschwindigkeit des Lichtes und der Körperdichte. (On the relations among velocity of propagation of light and the density of solids). *Widem. Ann. Phys.,* IX: 641–65.

Lorenz, L. (1880). Ueber die Refractionsconstante. (On refraction constants). *Widem. Ann. Phys.,* XI: 70–103.

Mandarino, J. A. (1976). The Gladstone–Dale relationship. Part I. Derivation of new constants. *Can. Min.,* 14: 498–502.

(1978). The Gladstone–Dale relationship. Part II. Trends among constants. *Can. Min.,* 16: 169–74.

(1979). The Gladstone–Dale relationship. Part III. Some general applications. *Can. Min.,* 17: 71–76.

(1981). The Gladstone–Dale relationship. Part IV. The compatibility concept and its application. *Can. Min.,* 19: 441–50.

Morse, S. A. (1968). Revised dispersion method for plagioclase. *Am. Mineral.,* 53: 105–16.

Nassau, K. (1983). *The physics and chemistry of color.* Wiley, New York.

Pauling, L. (1927). The theoretical prediction of the physical properties of many-electron atoms and ions. Mole refraction, diamagnetic susceptibility, and extension in space. *Proc. Roy. Soc. (London) A,* 114: 181–211.

(1960). *The nature of the chemical bond,* 3d ed. Cornell Univ. Press, Ithaca, N.Y., pp. 605–10.

Roberts, S. (1949). Dielectric constants and polarizabilities of ions in simple crystals and barium titanate. *Phys. Rev.,* 76: 1215.

Smith, J. W. (1955). *Electron dipole moments.* Butterworth, London.

Smyth, C. P. (1955). *Dielectric constant and molecular structure.* McGraw-Hill, New York.

Tessman, J. R., Kahn, A. H., and Shockley, W. (1953). Electronic polarizabilities of ions in crystals. *Phys. Rev.,* 92: 890–5.

Tilley, C. E. (1922). Density, refractivity and composition relations of some natural glasses. *Min. Mag.,* 19: 275–94.

Tröger, W. E. (1982). *Optische Bestimmung der gesteinbildenen Minerale, 5. Auflage. Teil 1.* Bestimmungstabellen von Profs. Bambauer, H. U., Taborszky, F., und Trochim, H. D. E. Schweizerbart'sche Verlagsbuchhandlung, Stuttgart.

Van Vleck, J. H. (1932). *The theory of electric and magnetic susceptibilities.* Oxford Univ. Press, London.

Winchell, A. N., and Winchell, H. (1961). *Elements of optical mineralogy,* 4th ed. Wiley, New York; Chapman and Hall, London.

Appendix: selected representative crystal structures

Figure 12.6. (A) The structure of zircon viewed along an a axis, showing the edge and corner linkages of ZrO_8 and SiO_4 polyhedra, and their translation along an a axis by axial glide planes a (001). (B) Projection of the SiO_4 tetrahedra of zircon along the c axis, showing the relation of all such tetrahedra to mirror planes (m), diamond glides (d), and 4_1 and 4_3 screw axes. These combine to yield space group $I4_1/amd$.

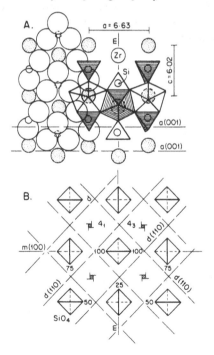

Illustrations are from H. W. Jaffe, *Crystal Chemistry and Refractivity* (New York: Cambridge University Press, 1988) and are numbered here as they are in that book.

Figure 12.8. A part of the garnet structure viewed along an a_3 axis. Unit cell is shown centered on a 4_3 left-handed screw axis. Many polyhedra are omitted for clarity. Elevations show that both AlO_6 octahedra and CaO_8 triangular dodecahedra rotate counterclockwise around the central 4_3 screw axis of the drawing, and translate polyhedra one-fourth the length of c with each 90° rotation.

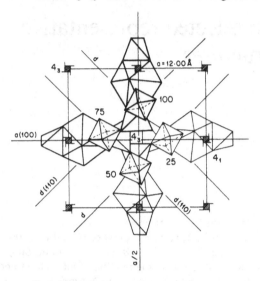

Figure 12.10. Packing model (upper) and polyhedral model (lower) of sillimanite, Al^{IV}—O—$Si[Al^{VI}O_4]$, projected along c, (001) projection. Al^{IV}—O^{II}—Si^{IV} bow ties connect with their corners to $[Al^{VI}O_4]$ chains running parallel to c, up toward viewer.

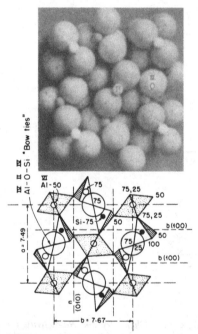

Figure 12.11. (A) Polyhedral projection on (001) of andalusite, Al^VO^{III}-$SI^{IV}[Al^{VI}O_4]$. Bow ties of sillimanite structure are replaced by paired, irregular, five-coordinated triangular dipyramids containing only Al. These corners connect to $[Al^{VI}O_4]$ chains along c. (B) Same projection as in (A) (001), showing the locations of large voids in the oxygen-packing assemblage. These result from the open packing of the paired CN V polyhedra, and their filling by amorphous to poorly crystalline matter gives rise to the chiastolite crosses that characterize andalusite in many occurrences.

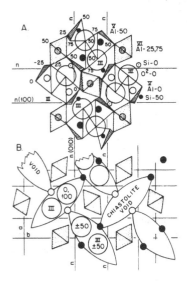

Figure 13.1. Packing model (upper) and polyhedral drawing (lower) of lawsonite, $4/CaAl_2(OH)_2[Si_2O_7] \cdot H_2O$ with $[Si_2O_7]$ tetrahedral groups rotated on c axis (turned toward viewer) for clarity. Note arrows locating orientation of H^+ ions. Plane of projection is (010).

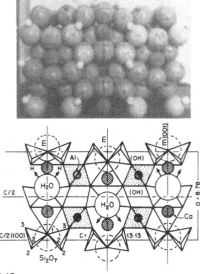

149

Figure 14.2. Polyhedral drawing of the structure of beryl, 2/□Al$_2$Be$_3$-[Si$_6$①$_6$②$_{12}$], (0001) projection.

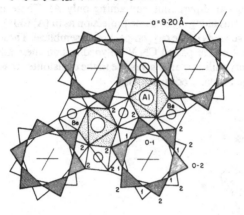

Figure 15.1. Similarities in the structure of phlogopite (A) and muscovite (B) (projected along a) with diopside (C) and pargasite (D) (projected along c). Amphiboles contain structural elements of both pyroxenes and micas.

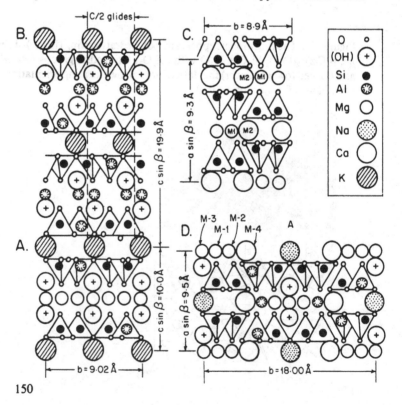

Figure 16.7. Packing model of sanidine projected along $a \sin \beta$, (100) projection, showing K^+ and $O-A^2$ ions on mirror plane (010) and tunnels along $a \sin \beta$.

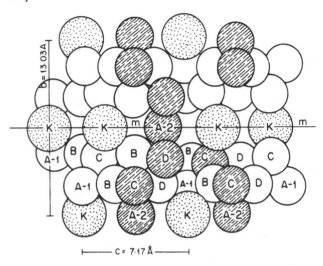

Figure 17.4. Unit cell projection (001) of ludwigite, showing numbering of anions, O-(1)–O-(5) (upper), and EBS distribution around each anion (lower).

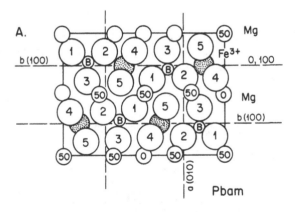

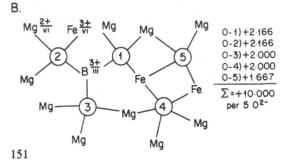

O-1)+2·166
O-2)+2·166
O-3)+2·000
O-4)+2·000
O-5)+1·667

$\Sigma = +10·000$
per 5 O^{2-}

Figure 18.9. (A) Polyhedral (octahedral) drawing of rutile, showing linkage of staggered rows of Ti octahedral chains parallel to *c*. (B) Polyhedral model of one unit cell of rutile is built around one octahedron.

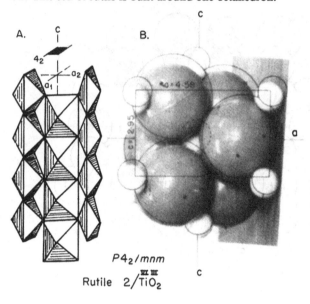

A.

B.

$P4_2/mnm$

Rutile $2/\frac{\overline{m}\,\overline{m}}{TiO_2}$

Figure 18.23. Diagrammatic representation of symmetry elements of the spinel structure $Fd3m$. Mg ions occupy the positions of C atoms of diamond. Note rotation of Mg—O tetrahedra and Al ions around 4_1 and 4_3 screw axes.

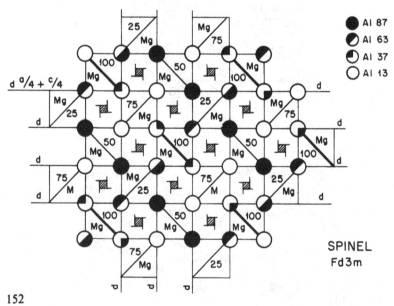

SPINEL
Fd3m

152

Mineral index

(Minerals for which essential crystal chemical data are given are paginated in **boldface.**)

acmite, **204**
adamite, 164
åkermanite, 115, **180–3**
alabandite, 104, **268**
albite, 79, 111, 137, **237–40**
alfordite, 312
almandine, 137, **158**
amphibole group, **215–21**
analcime, 80, 111, 127
anatase, 79, 115, 137, **280–2**
andalusite, 79, 105, 111, 115, 125, 137, **162–5**
  chiastolite var., 165
andesine, 239
andradite, 115, 138, **158**
anhydrite, 79, 141, **306–8**
anorthite, 79, 111, 115, 137, **237–41**
anthophyllite, 79–80, 111, 129, 148, **218–19**
antigorite, **227–8**
apatite, 79, **310–12**
aragonite, 20, 115, 137, **313–14**
armalcolite, **284**
augite, 79, 111, **204**
axinite, 126
azurite, **317–19**

baddeleyite, **274–6**
barite, 80, 98
bastnaesite, 65–6, 79, 140, **315–17**
belovite, 312
benitoite, 79
berlinite, 46
beryl, 64, 102, 111, 126, **187–9**
berzeliite, **158**
biotite, 228
bixbyite, **273–5**
borax, 137, **262**

britholite, 312
bromellite, 50, **289**
bronzite, **212–13**
brookite, 115, 137, **281–3**
brucite, **141–2**, 227
buddingtonite, 237
bunsenite, **104, 268**
bytownite, 79

cahnite, 252
calcite, 46, 79, 115, 121, 140–1, **312–14**
carbonate–apatite, 312
carnotite, 40
carobbiite, **268**
cassiterite, 137, **280**
celsian, 237
cerargyrite, 104
chlorapatite, 312
chlorite, 127, **142, 226, 228–9**
chloritoid, 126, **170–3**
chromite, 91–2, **296–8**
chrysoberyl, 80
clay mineral group, **228–9**
clinochrysotile, 229
clinopyroxene, 126, **204–8**
coesite, 137
colemanite, **252, 261**
columbite, **293–5**
cooperite, 35–40, 95, **289–90**
cordierite, 11, 126, **189–91**
corundum, 49, 79, 111–12, 137, **267–71**
cristobalite, **233–5**
cryolithionite, **158**
cubanite, 80
cummingtonite, **218–21**
cuprite, 40, 80, **301–2**

Subject index

158   *Subject index*

exsolution lamellae, augite host, 209–13
  C-pigeonite, 210
  hypersthene, 212
  P-pigeonite, 210
exsolution lamellae, orthopyroxene host,
  211–14
  augite, 211–13
  calcic plagioclase, 212–13
  garnet, 212
  ilmenite, 213
external symmetry, 71–5

face-centered lattice, F, 37
$Fd3m$ symmetry, 112
feldspars, 235–49
  Al—Si ordering, 239–44
  density, 237
  exsolution, 244–9
  iridescence, 247–9
  lattice parameters, 237
  optical properties, 237
  solid solution, 237–9
  structure, 241–4
  T-site ordering, 240–1
  voids, 242–3
$Fm3m$ symmetry, 112
formula weight, 108

garnet group
  CN sites, 156–9
  electrostatic bond strength, 62–3
  optical properties, 158
  screw axes, 159
  synthetic analogues, 105, 158
  YAG, YIG, 105, 156–9
Gladstone and Dale formula, 108, 122
glide planes, 78–81, 112
Goldschmidt's principles, 101
graphite
  comparison with diamond, 1–2, 35–6
  σ, π, sp^2 hybrid bonds, 35–6, 95
  van der Waals bonds, 36
ground (neutral) state, 15
gypsum plate, 122

half-life, 5
halite structure, 73, 112, 267–8
halogens, 10–11
Hermann–Mauguin symbols, 75
hexagonal close packing, HCP, 47–9
heteropolar (ionic) bond, 42
high energy d orbitals, 89–94
homopolar (covalent) bond, 42
Huttenlocher lamellae, 246–9
hybrid bond orbitals, 30–1, 95
  d^2sp^3 (pyrite), 36–8, 95
  dsp^2 (cooperite), 39, 40, 95
  sp (cuprite), 40
  sp^2 (graphite), 35–6, 95
  sp^3 (diamond), 35–6, 95

hydrogen bonds, 142
hydrogen burning, 11

I-beam structure, 208–9
$Im3m$ symmetry, 112
independence of polyhedra, 45–6
index of refraction, 118, 120–1
  variation with density, 131–2
indicatrix, optical, 121
interatomic distance, 42
interference figures, optical, 121–2
internal symmetry, 74–87
interplanar spacing, 76
inversion axes
  $\bar{3}$ calcite, 313
  $\bar{4}$ åkermanite, 181–2
  $\bar{4}$ SiF$_4$, 300
  $\bar{6}$ bastnaesite, 316
  $\bar{6}$ wadeite, 192
ionic bond, 22, 42–69
ionic bonding % vs. Δ_x, 43
ionic potential, 18–21
ionic radii, 52–8
ionic resonance energy, Δ, 23–4
ionization potential, 18
inosilicates, 149, 201–22
inverted pigeonite, 211–12
iridescence, 249
irrational planes, 211
isobars, 3–4
isoelectronic series, 18
isogyre, 122
isomorphism, 100–3
isostructural, 63, 100–4
isotones, 3–4
isotopes, 3–4
isotropy, 120–1

Jahn–Teller distortion, 90–4

K–Ar decay, 5
k-constants of oxides, 124–9
  general use, Jaffe, 124–5
  general use, Mandarino, 129
  rock-forming minerals, Jaffe, 126–8
kyanite-like zone, 170

lanthanides, 2, 10–11, 315
larvikite, 249
Lewis notation, 27–9
ligand, 89–90
lone pair orbitals, H$_2$O, 34–5
Lorentz–Lorenz equation, 136–8

M-site, feldspars, 238
M-1, M-2 site
  olivine, 95–7
  pyroxene, 206–8
M-1, -2, -3, -4 sites
  amphibole, 217
  ludwigite, 258–9

polar mineral structures (*cont.*)
  tourmaline, 199
  wurtzite, 288
  zincite, 286
polarizability, 18, 134–43
polarizability constants
  electronic, ions, 139
  molecular, minerals, 137
  molecular, oxides, 138
polarizer, 118
polyhedral sharing, 45–6
polyhedra with a common corner, 47
polarizing power, 118
polymerization, 45, 148–9, 254
polymorphism
  $Al_2O[SiO_4]$, 166–9
  C, 29–36
  $CaCO_3$, 312–14
  Fe_2O_3, 102, 271, 298
  $KAlSi_3O_8$, 239–41
  Mg_2SiO_4, 81, 97, 177, 296
  $NaAlSi_3O_8$, 237–40
  SiO_2, 233–5
  TiO_2, 276–83
  ZnS, 31–2, 288
primitive lattice, *P*, 37
pyrite
  chemical bonding, 36–9
  physical properties, 37–8
  structure, 38
pyroxene-like segment, 209
pyroxenes
  Al-rich orthopyroxene, 212
  amphibole unit cell comparison, 216
  cleavage, I-beam structure, 208–9
  exsolution, 209–14
  M-1, M-2 sites, 204–6
  optical properties, 204–5, 215
  unit cell data, 204–5, 216

quantum number
  angular momentum, 7
  azimuthal, 7
  magnetic, 7
  principal, 7
  spin, 7
quartz, handedness, 235–6

radiant visible energy, 15–18
radiation damage, 156
radioactive heat, 5
radioactivity, 5, 103, 155, 272
radiogenic substitution, 103
radius ratio, 42, 47–52
  coordination number, 47–51
rare earth garnets, 105
rare or noble gas, 10–11
Rb–Sr decay, 5
reconstructive phase transformation, 233
refraction, 119
refractivity variation
  coordination, 130
  covalency, 131
  density, 131–3
  electronegativity, 131
  valency, 131
resonance, 44
retardation, 119, 122
rotation axes
  crystal classes, 75
  crystals, 73
  symbols, 78
rotation axes, 2-, 3-, 4-, 6-fold
  2, alpha quartz, 231
  2, bastnaesite, 316
  2, sanidine, 238, 241
  3, calcite, 314
  3, CCP assemblages, 48–9
  3, HCP assemblages, 48–9
  3, pyrargyrite, 323
  3, pyrite, 37
  4, halite, 73–4, 112–13
  6, beryl, 188

sawhorse penetration twin, 169
scattering of X-radiation, 74
Schrödinger electron, 6
screw axes
  symbols, 78
  translations, 77
screw axes, 2_1
  åkermanite, 182
  azurite, 318
  datolite, 161
  epidote, 183–5
  sanidine, 238, 242, 243
  wadeite, 193
screw axes, 3_1, 3_2
  quartz, 235–6
screw axes, 4_1, 4_3
  anatase, 281
  diamond, 31, 112
  garnet, 159
  scheelite, 309
  spinel, 299
  zircon, 155
screw axes, 4_2
  cooperite, 290
  gold, 112
  halite, 112
  kamacite, 112
  PtO, 40
  rutile, 278–9
screw axes, 6_3
  apatite, 311
  niccolite, 322
  wadeite, 192
  zincite, 287–8
shared-edge shortening, 46, 96–9, 156, 189, 206–7, 269–70, 276–80

Printed in the United States
By Bookmasters